Dictionary of Space Technology

Test firing of a Space Shuttle main engine. [NASA]

Dictionary of Space Technology

Mark Williamson

BSc CPhys MInstP CEng MIEE FBIS

Space Technology Consultant

Adam Hilger
Bristol and New York

British Library Cataloguing in Publication Data

Williamson, Mark
  Dictionary of Space Technology
  1. Space sciences
  I. Title
  500.5

  ISBN 0-85274-339-4

Library of Congress Cataloging-in-Publication Data

Williamson, Mark.
  A dictionary of space technology / by Mark Williamson.
    401p. 24cm.
  ISBN 0-85274-339-4
  1. Astronautics--Dictionaries. 2. Rocketry--Dictionaries.
  I. Title.
  TL788.W55 1990
  629.4'03--dc20   89-19798

Cover photograph: Space Shuttle recovery of the Syncom IV-3 communications satellite on mission STS 51-I. [NASA]

Published under the Adam Hilger imprint by IOP Publishing Ltd
Techno House, Redcliffe Way, Bristol BS1 6NX, England
335 East 45th Street, New York, NY 10017-3483, USA

Typeset by CGS Print Group, Ellesmere Port
Printed in Great Britain by Dotesios Printers Ltd, Trowbridge, Wiltshire

To my parents
for allowing me the freedom to shape my own destiny

Contents

Preface and user's guide

Space technology covers a great many aspects of science and technology and, since its birth in the late 1950s, the subject has changed almost beyond recognition. This dictionary has been compiled with the aim of providing a convenient source of reference to some of the most important aspects of this developing technology.

One of the most difficult aspects of writing such a book is deciding what to put in and what to leave out. The basic premise was that it should be a *dictionary* rather than an *encyclopedia*. In other words, the emphasis is on defining the meaning of a word or phrase, and in some cases giving notes on its derivation and usage, rather than providing large amounts of background information. The entries have been written with the intention of enhancing the understanding of the subject, both for the practising specialist and the interested layman. To this end, the dictionary includes terms at all levels of sophistication: from the academic to the trivial; from the precision of engineering terminology to the frivolity of jargon.

The entries in this dictionary are arranged in alphabetical order. To assist the reader in research on a given topic, related entries are highlighted in the text and other important entries are included as a 'see also...'. Although every effort has been made to make the cross-references self-consistent, there may be cases where useful references have been omitted. Hopefully they are few and far between.

In an attempt to cover the fundamentals of space technology, the dictionary includes material which can be placed under the following headings:

Spacecraft technology
Communications technology
Propulsion technology
Launch vehicles
Manned spaceflight

Space Shuttle
Space centres and organisations
Orbits
Propellants
Materials
Physics and astronomy
Miscellaneous

To facilitate the use of the dictionary, a classified list of all the dictionary entries under these headings is included at the end of the book. Due to limitations on space and the intended nature of the book, the number of entries on specific spacecraft and space missions has been kept to a minimum.

A degree of judgement is implicit in the compilation of such a book, not only in what to include but also in the relative importance of the topics. It is inevitable that some readers will disagree with some of the entries, especially where preferred usage is under discussion. Where terms mean different things to different people, I have attempted to make this plain but, for the sake of the novice, have indicated the most common usage. New techniques and the new terminology to accompany them are being developed continually and no book can hope to cover them all. For this reason, this dictionary concentrates on the fundamental terms which will remain in common usage for the foreseeable future, adding a selection of historical and highly specific entries to add colour and depth to the subject. Any reader who has any suggestions for additions or improvements is invited to contact the author through IOP Publishing Ltd.

Mark Williamson
Arlesey, Bedfordshire
January 1989

Acknowledgments

Space technology draws heavily on a number of classical scientific disciplines including physics, chemistry, astronomy and, with the inclusion of manned spaceflight, biology. It is, however, the applications of these fundamental subjects that have expanded the field of space technology to the multidisciplinary subject it is today. Thus it includes aspects as diverse as orbital dynamics, communications, propellants and materials technology.

For this reason, I obtained the assistance of a number of friends and colleagues who specialise in one or other of these fields and asked them to read sections of the original manuscript covering their speciality. I am indebted to them for the time they gave so freely and would like to acknowledge them here. Although their contributions have improved the book, they are not of course accountable for the text as it appears — I have assumed editorial control and any errors are entirely my own responsibility!

I would particularly like to thank Alan Hutchinson, Ron Jones, Ron Cooper, Jehangir Pocha, Karen Burt, Geoff Statham and Dr W Berry whose contributions were invaluable. I am also grateful to Tom Keates, Anthony Giles, Ian White and Paul Brooks for their helpful comments. Many thanks are also due to Claude Bonnet, Richard Barnett and Clive Simpson.

I would also like to thank my wife, Rita, for her helpful comments on the manuscript and her considerable efforts in putting the dictionary into alphabetical order. And finally, thanks to Alan Burkitt for suggesting I write a dictionary of space technology in the first place.

MW

Dictionary of Space Technology

A

A-4

The original engineering designation for the V-2. See **V-2.**

ABLATION

The erosion of a surface, usually of a spacecraft **heat shield** on **re-entry,** due to friction with the molecular constituents of the atmosphere. A surface designed to ablate in such circumstances is known as an 'ablative surface' or 'ablative coating'.

ABM

An acronym for **apogee boost motor.** See **apogee kick motor (AKM).**

ABORT

The termination of a space **flight** or **mission** (used both as a verb and a noun).

As an illustration, the American **Space Shuttle** has four abort alternatives which can be used during the launch phase in the event of a **Space Shuttle main engine** failure:

1. Return To Launch Site (RTLS). Used between separation of the **solid rocket boosters** (SRBs) and the time when the next alternative (AOA) becomes available. The Shuttle flies on to burn the remaining **propellant**, turns back towards the **launch site**, jettisons the **external tank** and glides back to the launch site runway for a landing.

2. Abort once around (AOA). Used between about 2 minutes after SRB separation to when the next alternative (ATO) becomes available. The vehicle attains a **sub-orbital** trajectory, circles the Earth once and returns to the launch site.

3. Abort to orbit (ATO). Used when the Shuttle has passed the AOA-point and can attain orbit, although possibly of lower altitude than intended. In this case the **de-orbit**, **re-entry** and landing would be similar to a normal mission.

4. Transatlantic landings (TAL). An additional abort alternative which overlaps the latter part of the RTLS option, added when emergency runways become available on the eastern side of the Atlantic.

ABSOLUTE TEMPERATURE

Temperature measured on the 'absolute scale'. See **kelvin (K)**.

ABSOLUTE ZERO

The lowest temperature theoretically attainable; the temperature at which all molecular motion is zero (– 273.15 °C). See **kelvin (K)**.

AC POWER

A source of electrical power supplied with alternating current. See **power**. The great majority of spacecraft power systems use direct current (DC), AC being used only for special applications in scientific satellites and spacecraft payloads. The Hubble Space Telescope, for instance, uses a 20 kHz AC supply.

ACCELERATION DUE TO GRAVITY

See **gravity**.

ACCELEROMETER

A device for measuring acceleration.
[See also **inertial platform**.]

ACCESS ARM

A projection from a launch vehicle **service structure** which can be rotated towards the vehicle for access, either for general maintenance or the loading of a **crew**. If the arm is swung away from the vehicle at the moment of **launch**, usually when it carries **umbilical**s for **propellant** or electrical services, it is also known as a '**swing-arm**'.
[See also **white room**.]

ACCESS PANEL

See **closure panel**.

ACCESS TOWER

See **service structure**.

ACOUSTIC TEST CHAMBER

A ground-based test facility which simulates the acoustic environment experienced by a spacecraft during **launch.**
[See also **thermal-vacuum chamber, vibration facility, anechoic chamber.**]

ACTIVATED CHARCOAL CANISTER

See **environmental control and life support system (ECLSS).**

ACTIVE

A term applied to any device or system involved in mechanical or electrical action, or capable of a productive reaction to external stimuli; the opposite of **passive**. For example, an **amplifier** is an active device in a **communications system**: it makes an active contribution to the input **signal**. Similarly, a **heater** is an active device in a **thermal control subsystem**.

ACTUATOR

Any device which produces a mechanical action or motion; a servomechanism that supplies the energy for the operation of other mechanisms.
   Spacecraft actuators forming part of the **attitude and orbital control system** include **reaction control thruster**s, **reaction wheel**s and **momentum wheel**s.
[See also **nutation damper, solar sailing, orbital control, attitude control.**]

AERIAL

See **antenna**.

AERODYNAMIC HEATING

An increase in the **skin temperature** of a vehicle due to air friction, particularly at supersonic or hypersonic speeds [see **mach number**]. Sometimes called 'kinetic heating' although this can be caused by other forms of friction due to motion.
[See also **re-entry**.]

AERODYNAMIC STRESS

A generic term for the forces to which a **launch vehicle** or **spacecraft**, etc, is subjected during its passage through an atmosphere (during **launch**, **re-entry**, etc). See **aerodynamics**.

AERODYNAMICS

The study of air flow over a body and the resultant aerodynamic forces. See **lift**, **drag**, **thrust**.
[See also **aerospace vehicle**, **lifting body**, **lifting surface**, **fairing**, **dynamic pressure**.]

AEROSPACE VEHICLE

A term used for a space vehicle which, as part of its **launch** or **re-entry phase**, **is also capable of flight in the atmosphere using a lifting surface** (namely a wing) as well as conventional rocket propulsion.
[See also **spacecraft**, **single stage to orbit**, **HOTOL**.]

AEROSPIKE

See **plug nozzle**.

AEROZINE-50

A **liquid propellant** comprising 50% **hydrazine** and 50% **UDMH** (unsymmetrical dimethylhydrazine).
[See also **liquid propellant**.]

AGENA

A rocket **stage** used with the Atlas **launch vehicle**. See **Atlas**.

AIRBORNE SUPPORT EQUIPMENT (ASE)

Any equipment flown on a **spacecraft** or **launch vehicle** which supports a **payload** (physically or in terms of electrical supplies, etc). For example, a **cradle** (and its associated systems) which supports a spacecraft in the **payload bay** of the American **Space Shuttle**, or a Spacelab **pallet**.
[See also **flight hardware**, **ground support equipment (GSE)**.]

AIRFRAME

The supporting structure and aerodynamic components of a **launch vehicle**. The term, borrowed from aviation technology, tends to be applied only to space vehicles which have some contact with the

Earth's atmosphere during their launch phase, and not to **satellite**s and other similar **spacecraft**.
[See also **thrust structure, inter-tank structure, inter-stage, fairing, fin, skin, longeron, stringer, ogive, skirt, shroud, SYLDA, SPELDA, SPELTRA**.]

AIRLOCK

An airtight chamber which allows **astronaut**s and/or equipment to leave and/or enter a **spacecraft** without depressurising the entire vehicle. Early space **capsule**s were far too small to include a separate airlock so any **extra-vehicular activity (EVA)** required all crewmembers to don their **spacesuit**s before the capsule was depressurised. However, the American Space Shuttle **orbiter** has a removable airlock which can be installed in one of three different positions dependent on the mission: inside the crew compartment, allowing maximum use of the **payload bay**; inside the payload bay attached to the aft cabin bulkhead; or on top of the pressurised 'tunnel adapter' which links a **Spacelab** payload to the orbiter **cabin**. Two spacesuits are stored in the airlock and, during **EVA**, it can supply oxygen, cooling water, electrical power and communications services to the suited astronauts.

AKM

See **apogee kick motor**.

ALBEDO

The ratio of the intensity of light reflected from a body to that received from the Sun (in the 'visible spectrum' unless otherwise specified).

The fraction of the incident solar radiation returned to space by reflection from a planetary surface (solid or gaseous) is called 'planetary albedo', the average value of which for Earth, for example, is 0.34. In contrast, the *thermal energy* re-radiated by the Earth is known as 'earthshine'. Although important to the thermal design of spacecraft in **low Earth orbit**s, albedo and earthshine are only significant for geostationary spacecraft carrying devices at **cryogenic** temperatures.
[See also **thermal control subsystem, geostationary orbit (GEO)**.]

ALSEP

An acronym for Apollo lunar surface experiments package. The ALSEP, carried on all **Apollo** missions except Apollo 11, was stored in the **descent stage** of the **lunar module** and powered by a plutonium-238 **radioisotope thermoelectric generator (RTG)**, designated SNAP-27 (an acronym for systems of nuclear auxiliary power). The package contained **seismometers**, a **magnetometer**, and **solar wind** and lunar heat flow experiments.

ALTITUDE

(i) The vertical height of a body above the surface of a planet (typically above sea level for Earth).

(ii) In astronomy, navigation, etc, a measure of the angle above the horizon. See **elevation (angle)**.

ALTITUDE – AZIMUTH MOUNT

A structure for the support and guidance of an astronomical telescope or a satellite **earth station** which uses the 'horizon system' of celestial coordinates—see **azimuth**. In satellite applications it is referred to as an **elevation**-over-azimuth mount if its lower axis is perpendicular to the ground, and 'X–Y' if its lower axis is parallel to the ground. The term 'Az–El mount' is also sometimes heard. The major alternative mount for telescopes is the **equatorial mount**, known as a **polar mount** for earth stations.

ALUMINIUM (Al)

A low-density metal, widely used (when alloyed with other metals) in the aerospace industry. Historically, aluminium was alloyed with only a few elements close to it in the periodic table: magnesium, zinc, copper, silicon, manganese and lithium. However, new techniques, including rapid solidification technology, have trebled this number.

   Typical applications: spacecraft body-panel **face-skins**, mounting brackets and fittings (machined), launch vehicle adapter rings (forged). [See also **honeycomb panel, materials**.]

AM

See **amplitude modulation**.

AMES RESEARCH CENTER

See **NASA**.

AMMONIUM PERCHLORATE (NH_4ClO_4)

A solid **oxidiser** used in **rocket motors**. See **solid propellant**.

AMPLIFIER

An electrical device which increases the strength of an input signal and

presents a magnified replica of the signal at the output.
[See also **amplifier chain, high-power amplifier (HPA), low-noise amplifier (LNA), solid state power amplifier (SSPA), travelling wave tube amplifier (TWTA).**]

AMPLIFIER CHAIN

A general term for a number of amplifiers, and associated hardware, linked together in series. In a practical amplification device (for instance, a spacecraft **transponder**) a number of discrete, specialised amplifiers (e.g. pre-amplifiers, **low-noise amplifiers** and **IF amplifiers**) are commonly linked together to form a chain. Within the transponder one finds equipment divided, by function, into a **receive chain** and a **transmit chain**.

AMPLITUDE MODULATION (AM)

A transmission method using a modulated **carrier** wave, whereby the **amplitude** of the carrier is varied in accordance with the amplitude of the input signal; the **frequency** of the carrier remains unchanged.
[See also **modulation, frequency modulation (FM), phase modulation (PM), pulse code modulation (PCM), delta modulation (DM).**]

ANECHOIC CHAMBER

A ground-based test facility for the evaluation of **radio-frequency (RF)** equipment on a spacecraft, which simulates the RF propagation characteristics of **free space**. The walls and all service equipment and mounts are covered with RF absorbing material to reduce reflections (or 'echoes') to a minimum [see figure A1]. The larger chambers admit the whole spacecraft, but it is quite common to test only the **communications payload (antennas** and **transponders)** at an earlier stage in the design process.
[See also **thermal-vacuum chamber, acoustic test chamber, vibration facility**.]

ANNULAR NOZZLE

See **plug nozzle.**

ANOXIA

A lack of oxygen. See **hypoxia**.

Figure A1 The **European Space Agency**'s Orbital Test Satellite (OTS) in an **anechoic chamber**. [British Aerospace]

ANTENNA

The part of a radio system that enables a radio signal to be transmitted and/or received; the 'interface' between the radio equipment and the environment, between a '**free space**' RF wave and a guided wave. A radio **transmitter** 'excites' electric currents in the conductive surface layers of an antenna leading to the propagation of an electromagnetic wave; conversely, an incident radio wave 'excites' similar currents which are conducted to the **receiver**.

There are many different types of antenna, but using one method of categorisation four main types can be identified: wire, horn, reflector and array antennas [see figure A2]. For spacecraft applications, wire antennas operate chiefly at VHF and UHF frequencies, often taking the form of a helix, conical spiral or simple dipole. The other types operate mainly at **microwave** frequencies. Horn antennas are used by themselves on spacecraft to provide wide coverage of the Earth [see **global beam**], and as **feedhorn**s to illuminate reflector antennas. Both horns and reflectors are known as 'aperture antennas'. Another type of aperture antenna is the microwave lens, or 'dielectric lens', which, like an optical lens, can be designed to convert a spherical wave to a plane wave, thereby improving **directivity** [see **lens antenna**].

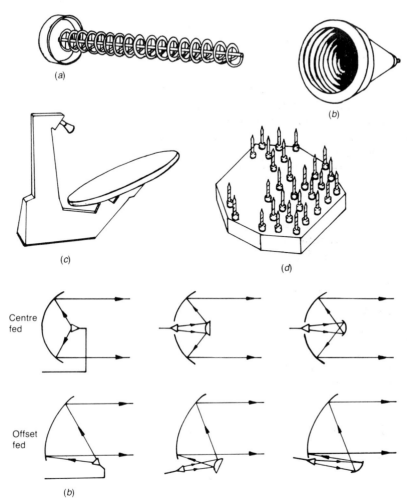

Centre
fed

Offset
fed

Figure A2 Antenna types and reflector–antenna configurations.
(a) Wire antenna (helix); (b) horn antenna (conical corrugated);
(c) reflector antenna (offset fed); (d) array antenna (TDRS phased
array helices). (e) Reflector antenna configurations; from left to
right: single reflector, Cassegrain, Gregorian.

The limited **gain** and relatively wide **beamwidth** of horn antennas has
led to the widespread use of reflector antennas, particularly on
communications satellites where high gain and narrow **spot beams** have
become increasingly desirable. Array antennas consist of a number of
radiating elements designed to act together to form a particular **beam**. The
array may comprise a number of slots in the wall of a **waveguide** (a 'slot-

array antenna'), a number of **dipole**s, helices, horns or reflectors, depending on the frequency, required beamwidth, etc.

The word 'aerial' is still used in the space industry (mainly by veteran British engineers), but is gradually being replaced by 'antenna'. Recommended plurals are 'antennas' for radio equipment, 'antennae' for insects.

[See also **frequency bands, communications payload, boresight, footprint, pointing, peak, axial ratio, *F/D* ratio, dipole, phased array, Cassegrain reflector, subreflector, antenna pointing mechanism, antenna radiation pattern, antenna platform, deployable antenna, unfurlable antenna, steerable antenna, elliptical antenna, parabolic antenna, isotropic antenna, omnidirectional antenna, TT&C antenna**.]

ANTENNA ARRAY

An assembly of **antenna**s, not necessarily electrically coupled as they are in an **array antenna**.
[See also **antenna farm, antenna platform, antenna module**.]

ANTENNA EFFICIENCY

See **aperture efficiency**.

ANTENNA FARM

A collective term for a number of **antenna**s. Usually refers to an orbiting **platform** carrying an assembly of communications antennas, but sometimes extended to the **antenna platform** of a much smaller **satellite** or a collection of antennas at an **earth station** [sense (ii)].
[See also **antenna module, antenna array**.]

ANTENNA FEED

See **feedhorn**.

ANTENNA GAIN

The ratio of the signal power at the **output** of an **antenna** to that received at the **input**, usually measured in decibels (dB). The majority of antennas used in **satellite communications** are parabolic dishes which, when used on the input side, increase the gain of the signal passed to the **receiver** and, when on the output side, increase the gain of the signal transmitted through space. This definition views the antenna as a gain-producing component in a **transmit** or **receive chain**; viewed in a more theoretical sense as an independent entity, the word 'gain' refers to an increase in signal power over and above that which would be available from an **isotropic antenna**.

Two expressions commonly used to calculate antenna gain are given below:

$$\text{Peak gain:} \quad 10\log_{10} \eta(\pi D/\lambda)^2$$

where η is antenna efficiency (%), D is diameter (m) and λ is wavelength (m); and

$$\text{Half-power gain:} \quad 10\log_{10} \eta(27\ 800/\Theta\Phi)$$

where η is antenna efficiency (%), and Θ and Φ are the orthogonal **half-power beamwidths** in degrees. Alternative values for the constant may be found in other texts, where a distinction is made between antennas of different types.
[See also **equivalent isotropic radiated power (EIRP)**, **power flux density (PFD)**, **directivity**, **decibel (dB)**.]

ANTENNA MODULE

A self-contained section of a modular spacecraft containing the communications **antenna** subsystem.
[See also **payload module**, **service module**.]

ANTENNA PLATFORM

(i) The panel of a satellite's body which supports the **antennas**. Usually the Earth-pointing face, although antennas can be mounted on other faces too [see **deployable antenna**].

(ii) Another name for **antenna farm**.

[See also **antenna module**, **antenna array**.]

ANTENNA POINTING MECHANISM (APM)

A device which gives a spacecraft **antenna** a finer degree of **pointing** control than that offered by the body of the spacecraft itself, particularly important for **spot beam** antennas.
[See also **attitude control**.]

ANTENNA RADIATION PATTERN

A measure of the directional sensitivity of an **antenna**; graphically, a plot of the radiated field strength against the angle from **boresight** [figure A3].

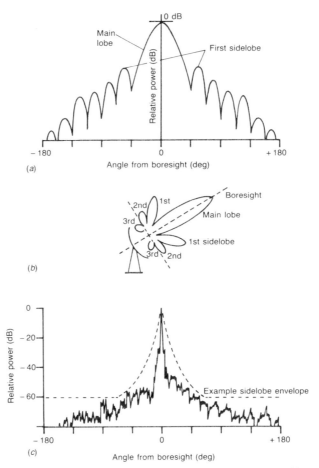

(a)

(b)

(c)

Figure A3 Antenna radiation patterns. (*a*) Conventional linear plot (simplified), (*b*) pattern in polar coordinates, (*c*) example pattern for a 3 m earth station.

The pattern is important since it determines the **beamwidth** and **directivity** of the antenna. Sensitivity in a direction outside the main **beam** or main 'lobe' of the pattern is known as a 'sidelobe'. This represents a detrimental attribute of an antenna, because it indicates that some of the power radiated from the antenna will not be contained within the main beam which, apart from being wasteful, could lead to **interference** with other systems. Equally, a signal received from the side-lobe direction could interfere with the system in question.

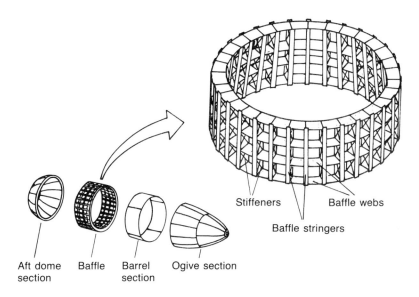

Figure A4 An **anti-slosh baffle** in a propellant tank.

ANTI-SLOSH BAFFLE

A structure in a **liquid propellant** tank which damps out the motion of the liquid known as '**sloshing**', which can disturb a vehicle's flight dynamics [see figure A4]. Various different structures are used (e.g. vanes, rings, truncated cones, etc). Some tanks also have anti-vortex baffles, which minimise the propellant's tendency to swirl as it flows out of the tank (like water down a plug-hole). Vortices can produce bubbles of gas which would otherwise pass to the **engine**(s) producing uneven combustion.

AOCS

See **attitude and orbital control system**.

AOS

An acronym for acquisition of signal. Typically used to denote the time a tracking **earth station** acquires a **spacecraft**'s **telemetry** signal during **launch** and **transfer orbit** phases. Also used when a **communications** link with a spacecraft has been re-established after a period of 'radio silence' (e.g. following blockage by a **planetary body**, a period of radio **interference** or during **re-entry** [see **S-band blackout**]). The opposite of 'loss of signal' (LOS).
[See also **signal**, **carrier**.]

AP

An acronym for **ammonium perchlorate (NH_4ClO_4)**.

APERTURE ANTENNA

A 'horn antenna' or a 'reflector antenna'. See **antenna**.

APERTURE BLOCKAGE

The reduction in the effective area of an **antenna** reflector by obstructions in the aperture, such as a **feed** or **subreflector** and their supports. Blockage decreases the **aperture efficiency** and degrades the **side-lobe** performance due to diffraction.

APERTURE EFFICIENCY

The efficiency of an '**aperture antenna**'; a term used in the calculation of **antenna gain**, etc, which quantifies how effectively the antenna uses the **RF power** transmitted or received.

  This can include 'illumination efficiency' or 'aperture taper efficiency' (the efficiency with which the power is distributed over the surface of an antenna reflector by the **feed**); the degree of **spillover**; the degree of **aperture blockage**; diffraction effects; phase errors; and **polarisation** and 'mismatch' losses.

APHELION

The furthest point from the Sun in an elliptical solar orbit (from the Greek for 'away from the sun'); the opposite of **perihelion**. See **apogee**.

APM

See **antenna pointing mechanism**.

APOAPSIS

The point in an **orbit** furthest from the centre of gravitational attraction; the opposite of **periapsis**. See **apogee, aphelion**.

APOGEE

The point at which a body orbiting the Earth, in an **elliptical orbit**, is at its greatest distance from the Earth; the opposite of **perigee**. The word is derived from the Greek 'apogaios': the prefix 'ap' meaning 'away from'; the suffix 'gee' referring to 'Earth'.
[See also **orbit, apogee kick motor**.]

APOGEE BOOST MOTOR (ABM)

See **apogee kick motor (AKM)**.

APOGEE ENGINE

See **liquid apogee engine**.

APOGEE KICK MOTOR (AKM)

A **rocket motor** used to transfer a satellite from **geostationary transfer orbit** (GTO) to **geostationary orbit** (GEO) [see figure A5]. Alternatively called an apogee boost motor (ABM); sometimes called an apogee stage.

The AKM is fired when the spacecraft reaches the point in the GTO ellipse furthest from the Earth, the **apogee**. This type of motor is an integral part of most geostationary satellites and is often installed inside a central thrust cylinder or other **thrust structure** [but see **inertial upper stage (IUS)**]. The AKM uses **solid propellant**, like the similar but larger **perigee kick motor (PKM)**; in a **liquid propellant** system, the equivalent device is known as a **liquid apogee engine** (LAE). [See also **rocket engine.**]

Figure A5 The MAGE 2 **apogee kick motor** used on the **Giotto** Comet Halley interceptor. [SEP]

APOGEE STAGE

A **launch vehicle** stage which injects a spacecraft into **geostationary** orbit by firing its **engine**(s) at the **apogee** of the **transfer orbit**. Sometimes used as an alternative term for **apogee kick motor**. [See also **stage**.]

APOLLO

A series of American **manned spacecraft** designed to take three **astronaut**s to lunar **orbit** and land two of them on the Moon. Apollo comprised two spacecraft: a command and service module (CSM) [see figure A6] and a lunar module (LM), formerly called the LEM (for lunar excursion module). After the third **stage** of the **Saturn V** launch vehicle had injected the Apollo combination onto a lunar trajectory [**trans-lunar injection (TLI)**], the CSM separated and withdrew the LM from its launch **fairing** on top of the S-IVB stage. The service module's main engine was used to place the spacecraft in lunar orbit, to return it to Earth and for major **mid-course correction**s. It used the **propellant**s UDMH and **nitrogen tetroxide (N_2O_4)** which were also employed by its four sets of RCS [**reaction control system**] thrusters, used for **rendezvous** and **docking**, minor **manoeuvre**s and as **ullage rocket**s.

Figure A6 Apollo 15 command and service module in lunar orbit, 30 July 1971. Note the **docking probe** on top of the command module. [NASA]

The command module, the only part of the whole vehicle to return to Earth, was the crew **cabin** for the outward and return journeys as well as the **re-entry** vehicle: it had an ablative **heat shield** made from an epoxy resin bonded to a stainless steel honeycomb. A tunnel in the nose gave access to the LM. The lunar module [figure A7] comprised a **descent stage** and an **ascent stage**.

Figure A7 Apollo 15 lunar module and **lunar roving vehicle** at the Hadley Rille landing site, 31 July 1971. **Astronaut** James Irwin salutes the Stars and Stripes. [NASA]

Following a number of unmanned launches of Apollo hardware, the first manned mission (Apollo 7/**Saturn 1B**, 11–22 October 1968) tested the CSM in Earth orbit; Apollo 8, which was launched by a Saturn V without a LM, made ten orbits of the Moon; Apollo 9, which conducted **docking** tests in Earth orbit, was the first manned flight with a complete Apollo spacecraft; and the Apollo 10 LM descended to within 15 240 m of the lunar surface in a 'dress rehearsal' for the landing. Apollo 11 made the first lunar landing in the Sea of Tranquility (0° 41′ 15″ N, 23° 26′ E) at 9.18 pm BST 20 July 1969, the first foot being placed on the surface at 3.56 am BST 21 July (21.8 kg rock returned) [figure A8]. Apollos 12, 14, 15, 16 and 17 also made lunar landings; the latter three carried a **lunar roving vehicle** (LRV) to extend their coverage of the lunar surface. Apollo 13 suffered an explosion of an oxygen tank in its service module en route to the Moon and failed to make a landing.

In summary of the Apollo missions, 12 men spent 160 man-hours on the Moon, collected over 2000 rock samples totalling 380 kg, and took over 30 000 photographs; 60 major experiments were placed on the lunar surface [see **ALSEP**] and 30 were carried out from lunar orbit.

Figure A8 Astronaut Edwin 'Buzz' Aldrin photographed by Neil Armstrong, the first man to set foot on the Moon, during the **Apollo** 11 mission, 16–24 July 1969. Note Armstrong and a leg of the Lunar Module 'Eagle' reflected in Aldrin's visor. [NASA]

APOLLO–SOYUZ TEST PROJECT

A **docking** mission between an American **Apollo** and a Soviet **Soyuz** spacecraft which developed from a political agreement between President Nixon and Premier Kosygin. The spacecraft docked on 17 July 1975, Apollo 18 carrying three **astronaut**s, and Soyuz 19 carrying two **cosmonaut**s (launched 2 days earlier).
[See also **docking module**.]

APPROACH AND LANDING TEST

See **Enterprise**.

APU

See **auxiliary power unit**.

AREA EXPANSION RATIO

See **expansion ratio**.

ARIANE

A European three-stage **launch vehicle** developed in the 1970s under
the auspices of the European Space Agency (**ESA**). It was proposed by
the French space agency after the cancellation of the **Europa** launch
vehicle programme and given the 'go-ahead' in 1973.

Ariane has evolved through several variants: Ariane 1, 2, 3 and 4 with
payload capabilities of 1850, 2175, 2700 and up to 4200 kg, respectively,
to **geostationary transfer orbit** (GTO). The vehicles use the propellants
nitrogen tetroxide and UDMH in the first and second stages and **liquid
oxygen/liquid hydrogen** in the third stage. Variants 1, 2 and 3 use the
same first and second stages; 2 and 3 have a 'stretched' third stage; and
3 uses two **strap-on** solid rocket boosters to augment the first stage
thrust [see figure A9]. Ariane 4 has a stretched first stage, but uses the
same second and third stages, although strengthened to withstand
increased launch **load**s. A number of solid and liquid **propellant** strap-
ons can be added to form six different configurations: no boosters; 2
solids; 4 solids; 2 liquids; 2 solids and 2 liquids; and 4 liquids.

Figure A9 Ariane 3 **launch vehicle**. Note strap-on **solid rocket
booster**s. [CEF, Bernard Paris]

The first flight of the Ariane from its launch site at the **Guiana Space Centre** was in December 1979. The majority of its payloads are commercial **communications satellite**s placed in **geostationary orbit**, although it can also launch spacecraft into **low Earth orbit** and **Sun-synchronous orbit**.
[See also **SYLDA**, **SPELDA**.]

ARRAY

See **solar array**, **antenna array**.

ARRAY ANTENNA

See **antenna**.

ARRAY BLANKET

See **solar array**.

ARRAY SHUNT REGULATOR

A device which switches sections of a spacecraft **solar array** on and off in accordance with the demand for **power**.
[See also **shunt dump regulator**.]

ASCENT ENGINE

A spacecraft **rocket engine** used for an ascent from the surface of a **planetary body**; the main engine in the spacecraft's 'ascent stage' (e.g. the ascent engine of the **Apollo** lunar module).
[See also **descent engine**.]

ASE

See **airborne support equipment**.

ASSEMBLY

(i) A blanket term for any group of components assembled to perform a particular function (e.g. bearing and power transfer assembly (**BAPTA**), **thruster** assembly, etc).

(ii) Construction (as in **vehicle assembly building** (**VAB**)).

ASTP

See **Apollo–Soyuz test project**.

ASTROMAST

A type of extendable truss structure used to **deploy** items, such as solar arrays and **deployable antenna**s, from spacecraft [figure A10]. See **solar array**.

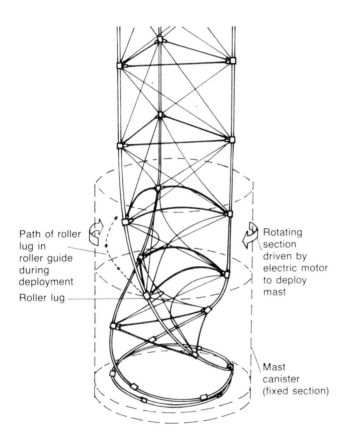

Path of roller lug in roller guide during deployment

Roller lug

Rotating section driven by electric motor to deploy mast

Mast canister (fixed section)

Figure A10 An **astromast** used to deploy a **solar array**.

ASTRONAUT

Before the introduction of the American **Space Shuttle**, the term was applied to any person who travelled in space [but see also **cosmonaut**], all of whom were pilots specifically trained for the duties implied by spaceflight. The only exception, in the history of American spaceflight (pre-shuttle), was 'scientist–astronaut' and **Apollo** 17 lunar module pilot, Harrison ('Jack') Schmitt, a geologist who became an astronaut. Now that any 'reasonably fit' person is physically capable of flight aboard the Shuttle, the term 'Astronaut' tends to be confined to a person who has trained as a 'career astronaut'. Other participants in a Shuttle flight are called **mission specialists** or **payload specialists**.

ASTRONAUT MANOEUVRING UNIT (AMU)

See **manned manoeuvring unit (MMU)**.

ASTRONAUTICS

Strictly, the study of the fundamental physical principles governing **spaceflight** (i.e. **orbit**s, trajectories, **gravitational perturbation**s, etc), although the word has come to be used as a blanket term for the science and technology of spaceflight, **space science** and the exploration of space. In either case, 'astronautics' has little to do with the 'antics of astronauts'!
[See also **cosmonautics**.]

ASTRONOMICAL SATELLITE

An orbiting spacecraft concerned with astronomical observations, usually based in **low Earth orbit**. The main advantage of astronomy from orbit is the lack of an atmosphere, the limiting factor for many branches of terrestrial astronomy.
[See also **space astronomy**, **space telescope**.]

ASTRONOMICAL UNIT (AU)

The mean distance between the Earth and the Sun (i.e. the radius of the Earth's orbit, 1.495×10^{11} m); a unit of distance used in astronomy.

ATLANTIS

The name given to the fourth '**flight model**' of the American Space Shuttle **orbiter** (orbiter vehicle OV-104), which was first launched on 3 October 1985 (STS 51-J). Atlantis was named after a sailing ship used for ocean research from 1930 to 1966.
[See also **Space Shuttle**, **space transportation system (STS)**, **Enterprise**, **Columbia**, **Challenger**, **Discovery**, **Endeavour**.]

ATLAS

An American **launch vehicle** developed from an intercontinental ballistic missile (**ICBM**), which had its first flight in June 1957 and became an operational weapon system in 1959. The Atlas was considered a 'one-and-a-half' **stage** vehicle. Its three main engines were ignited simultaneously at launch, but only one, known as the 'sustainer', remained attached to the propellant tanks for the full duration of the flight. The other two, called 'boosters', were jettisoned along with their **fairing**s, etc, part way through the flight. This configuration became the first stage of successive versions of the Atlas launch vehicle [figure A11] (**propellant: liquid oxygen/kerosene**).

Figure A11 The **launch** of an **Atlas** Centaur at the Eastern Test Range, Florida. Note the **swing-arm** on the **service structure** and the rocket's **vernier engine**. The white cloud around the base of the vehicle is composed of ice particles formed by the freezing effect of the rocket's **cryogenic propellant**. [General Dynamics]

The Atlas-D variant, for instance, was used to orbit the **Mercury** capsule; the Atlas–Able and Atlas–Agena two-stage vehicles launched the Ranger probes to the Moon, the early Mariners to Mars and Venus, etc; and the Atlas–Agena D carried the Agena target docking vehicle for the **Gemini** programme. The Centaur stage used on the Atlas–Centaur was America's first liquid oxygen/**liquid hydrogen** stage: it first flew in May 1962, and became **operational** in May 1966 with the launch of the Moon **probe** Surveyor.

The Atlas became a commercially operated launch vehicle in the late 1980s. The **payload** capability of the Atlas H is about 2000 kg to **low Earth orbit** (LEO), and that of the Atlas G/Centaur is about 6100 kg to LEO or 2350 kg into a **geostationary transfer orbit (GTO)** or about 1220 kg to **geostationary orbit** (GEO).
[See also **Delta, Titan, Scout.**]

ATMOSPHERIC ATTENUATION

The weakening of a radio **signal** due to its passage through a planet's atmosphere.

The attenuation of a **radio-frequency** signal in the Earth's atmosphere is of everyday importance for **satellite communications** and is used in the calculation of the communications **link budget.** It is partly due to oxygen and water vapour, whose molecules are excited into higher energy states by absorption of the radiation. Precipitation in its various forms also acts as a radiation absorber. The degree of attenuation is heavily dependent on the **frequency** of the radiation: attenuation increases with increasing frequency [see figure A12]. In a satellite communications system, the **elevation angle** of the **satellite** as measured from the **earth station** also has an effect: attenuation decreases with increasing elevation angle (since the atmospheric path is shortest at the zenith).

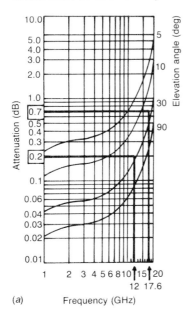 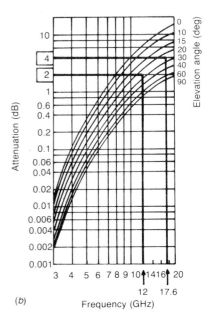

Figure A12 The variation of **atmospheric attenuation** with **frequency** and **elevation angle** (a) for oxygen and water vapour and (b) for rainfall (with example values for **DBS uplink** and **downlink** frequencies).

ATMOSPHERIC DRAG

The resistance to the motion of a body due to a planet's atmosphere.
Spacecraft in very **low Earth orbit**s can have their orbital velocity decreased to such an extent that they re-enter the atmosphere and 'burn up'. An increase in solar activity [i.e. a **solar flare**] tends to increase the average height of the atmosphere, thereby increasing the drag at a particular height. This can decrease the **lifetime** of spacecraft in LEO, particularly those with large surface areas (e.g. the USA's **Skylab** space station, which re-entered earlier than expected for this reason).
[See also **drag compensation**.]

ATTENUATION

(i) The function of an **attenuator**.

(ii) The natural weakening of a **signal** as it passes through an attenuating medium such as the atmosphere. See **atmospheric attenuation**.

ATTENUATOR

A device which provides a measured degree of **attenuation**, or weakening, of a signal. This may be necessary as a fine adjustment where standard equipment provides too great a degree of amplification, or to terminate the open end of a **transmission line** which can occur at a switch when power is routed along the alternative course. Attenuators can be of the fixed or variable type, and are colloquially known as 'pads'. The variable type on board **spacecraft** can be adjusted from the controlling **ground station**.

ATTITUDE

The orientation of a spacecraft's axes with reference to a number of known points, either on Earth or in space.
[See also **spacecraft axes, attitude control**.]

ATTITUDE AND ORBITAL CONTROL SYSTEM (AOCS)

A spacecraft subsystem which combines the functions of **attitude control** and **orbital control**.

ATTITUDE CONTROL

The ability to maintain or change a spacecraft's **attitude**; the subsystem or process by which this control is effected.
Attitude control is required to ensure that **sensor**s, communications **antenna**s and other instruments remain pointing in the right direction. Most spacecraft have an attitude control subsystem, since they need to

counter the perturbing forces experienced in orbit [see **perturbations**] and maintain a specified orientation. The system usually comprises a set of **reaction control thruster**s and may include other **actuators** such as **reaction wheel**s.

Attitude control is usually combined with **orbital control** in a common subsystem, the 'attitude and orbital control system' (AOCS). [See also **pointing**.]

AU

See **astronomical unit**.

AUTOPILOT

An automatic guidance and control system on a **launch vehicle**, which calculates the vehicle's **attitude** and position using an **inertial platform** and flys the rocket according to preprogrammed **data**. Usually part of the vehicle's **instrument unit**.

AUXILIARY POWER UNIT (APU)

One of three devices in the American Space Shuttle **orbiter** that provides hydraulic power for main engine steering ('**thrust vector** control'), landing gear **actuators**, brakes, nose-wheel steering, **elevon** and rudder/**speedbrake** control, etc. The three APUs (one to control each **Space Shuttle main engine**) are used during launch and for the approach and landing. The APU is fuelled by **hydrazine** which is decomposed in a **gas generator**; the gas is used to drive a turbine which drives a hydraulic pump.
[See also **power supply**, **fuel cell**.]

AVIONICS

Electronics technology applied to aeronautics (from *avi*(ation electr)*onics*), and by extension to **astronautics**; electronic equipment on board a **spacecraft**—usually a manned spacecraft by analogy with aviation. Avionics equipment can include guidance and navigation systems, **communications** equipment, and items such as **gyroscopes**, **radar altimeters**, etc.

AXIAL RATIO

A measure of **polarisation** circularity in a **cross-polarised** antenna. [See also **cross-polar discrimination**.]

AXIAL THRUSTER

A **reaction control thruster** on a **spin-stabilised** spacecraft mounted in alignment with the spacecraft's spin axis (colloquially and collectively termed 'axials'). It produces **thrust** in a direction parallel to the spin axis and is typically used for north—south station keeping and **precession** control [see **orbital control**].

[See also **radial thruster**.]

AZIMUTH

A measure of the angle between the observer's meridian and the meridian of the subject under observation, in the horizon system of celestial coordinates. See also **elevation (angle)**: knowledge of a **satellite**'s azimuth and elevation angle allow an **earth station** to be aligned with it. An alternative 'equatorial' or 'geographical' system uses **declination** and **hour-angle**. Azimuth is measured clockwise from due south for satellite applications, as in astronomy; it is measured clockwise from due north in navigation circles.

B

BACK-OFF (from SATURATION)

In **telecommunications**, the degree to which the output **power** of an **amplifier** is reduced from the point where it is described as 'saturated' to its operational level (it is said to be 'backed-off from saturation'); the process of reducing the input power level to a point where the amplifier is operating in its 'linear' region (i.e. where the output power bears a linear relationship to the input power). See **saturated output power**.

BACKGROUND NOISE

A colloquial term for '**noise**'; a blanket term for any undesired electrical disturbance in a circuit or communications system independent of its origin. See **thermal noise, broadband noise, shot noise**.

BACKHAUL LINK

A terrestrial telecommunications link from a satellite **ground station** to a local terrestrial switching centre, which channels the signal into the public network; also called a 'terrestrial tail'.

BACK-UP

A spare, substitute or **redundant** component, **system, spacecraft**, etc. [See also **prime system, prime spacecraft, ground spare, in-orbit spare**.]

BAFFLE

Any device which restrains the motion of a fluid or restricts the passage of light (or other radiation in the **electromagnetic spectrum**). For example, an **anti-slosh baffle** which restrains the motion of a **liquid propellant** in a **propellant tank**, or a 'light baffle' which reduces the

radiation from unwanted sources entering a **telescope** or other **imaging** device.
[See also **sunshield**.]

BAIKONUR COSMODROME

The main Soviet **launch site**, located near the town of Tyuratam to the east of the Aral Sea (at approximately 46°N, 63°E). Although the Soviets refer to it as the 'Baikonur Cosmodrome' it is some 300 km from the town of Baykonyr (alternative spelling) and only about 20 km from Tyuratam. The site was developed in 1955 to provide **ICBM** test facilities and has been used since the launch of **Vostok** 1 for manned and unmanned launches.
[See also **Northern Cosmodrome**.]

BALANCE MASS

The additional mass added to a completed spacecraft before **launch** to ensure its static and dynamic balance during all phases of its mission. For example, static balance is important for **attitude control** in ensuring that **thruster** firings, etc, have the desired effect; dynamic balance is important in reducing **nutation**, when a spacecraft is **spin-stabilised**.
[See also **mass budget**.]

BALLISTIC FLY-BY

See **ballistic trajectory**.

BALLISTIC TRAJECTORY

The path described by a body moving in three dimensions under the influence of gravitational forces, and any resistive forces due to the medium of travel (although ideally thrust, drag and lift would be zero); the **trajectory** followed by a body moving in accordance with the laws of ballistics, the study of projectile flight dynamics.

For example, a **launch vehicle** without **lifting surface**s travels along such a path once **propulsion** ceases (like a 'ballistic missile')—during **guidance** or propulsive phases the vehicle is not considered a ballistic body. A spacecraft using the gravitational field of a planet as its means of locomotion is undertaking a 'ballistic fly-by' (e.g. **Pioneer**, **Voyager**, etc).
[See also **free-return trajectory**, **trans-lunar trajectory**.]

BALLUTE

A combination of balloon and parachute sometimes used to assist in the

deceleration of a **spacecraft** prior to **touchdown**.
[See also **retro-rocket**.]

BAND (FREQUENCY)

See **frequency band**.

BAND-PASS FILTER

A **filter** with both high- and low-frequency cut-offs which allows a specified band of signal frequencies to pass. This is the generic term for a group of filters designed to confine a signal to a specified **bandwidth** (namely **input filter**, **output filter** and **channel filter**).

BANDWIDTH

The width of a **frequency band**, typically measured in MHz. The bandwidth is one of the factors which gives a measure of the amount of information (telephone calls, data streams, etc) that can be transmitted over a communications link, for instance through a satellite **transponder**.
[See also **capacity**.]

BAPTA (BEARING AND POWER TRANSFER ASSEMBLY)

A rotating interface between a spacecraft body and a **solar array**.

On a **three-axis-stabilised** spacecraft the BAPTA keeps the array pointing towards the Sun and transfers the power generated to the spacecraft equipment. Each array has its own BAPTA, which accepts commands from **sun sensor**s mounted on the arrays. On a **spin-stabilised** spacecraft the cylindrical array and the majority of the vehicle rotates while the **communications payload** is de-spun to allow the **antenna**s to point towards Earth—the single BAPTA transfers power to the payload.

Apart from the electronics, a typical BAPTA consists of a bearing, a drive motor and a slip-ring unit. The mechanical part of the device is known alternatively as a solar array drive mechanism (SADM) which, in combination with the solar array drive electronics (SADE), is called a solar array drive assembly (SADA).

BARBECUE ROLL

A slow **roll** given to a spacecraft to equalise the heating effect of the Sun over its surface; a **passive** method of thermal control which reduces the thermal gradient across the spacecraft. Used particularly for spacecraft on a **trajectory** between two astronomical bodies, where the Sun's direction is approximately constant (e.g. used for **Apollo** on a lunar trajectory: rotation rate about 0.1° s^{-1}, or one full turn per hour). [See also **thermal control subsystem**.]

BASEBAND

The **frequency band** that a signal occupies when initially generated; the signal **bandwidth** prior to **modulation** and subsequent to **demodulation**. The signal may be the output of an item of equipment, such as a telephone, telex, TV camera, computer, etc.

BATTERY

A source of **DC power** carried aboard most spacecraft as a **secondary power supply**. Batteries are required to store power for use during **eclipse** and during the **transfer orbit**, for satellites that have no array capability during this phase. The battery subsystem is often referred to as the secondary power supply for this reason. Two types of battery commonly used on satellites in Earth orbit are the nickel—cadmium (NiCd) cell and the nickel—hydrogen cell (NiH$_2$). [See also **depth of discharge (DOD)**, **duty cycle**, **solar array**.]

BATTERY RECONDITIONING

The controlled discharging and subsequent recharging of a **battery** conducted on a regular basis to improve battery performance. The batteries on satellites in **geostationary orbit** are reconditioned during the **solstice** period, when battery **power** is not required.

BAUD

The transmission or signalling rate on a **data** channel, where a baud is a signalling element per second. The data transfer rate (measured in bits per second) is usually a simple multiple of this. Named after the French inventor J M E Baudot. [See also **bit**, **bit rate**.]

BEACON

(i) A low-power **carrier** which may be unmodulated for **propagation** tests or modulated with **telemetry** or **tracking** data. See **modulation**.

(ii) A device on a **spacecraft** which transmits a signal that allows it to be tracked by a **ground station** or another spacecraft. Also used to conduct propagation tests which ascertain the effect of the atmosphere on the beacon signal.

[See also **tracking**, **atmospheric attenuation**.]

BEAM

(i) The term used to describe the geometrical distribution of **radio-frequency** radiation formed by a **satellite** communications **antenna** (e.g. if the antenna aperture is circular, the beam is conical).
[See also **beamwidth, global beam, hemispherical beam, zone beam, spot beam, multiple beam.**]

(ii) See **electron beam.**

(iii) See **beam-builder.**

BEAM AREA

An area on the surface of the Earth, defined for the purposes of **satellite communications**, which is within the **half-power beamwidth** of the spacecraft **antenna.**
   The periphery of the beam area corresponds to the − 3 dB points on the satellite **antenna radiation pattern** and is therefore bounded by the − 3 dB contour. In many cases the beam area practically coincides with the **coverage area**, but the two are defined from different premises and should not be confused.

BEAM-BUILDER

An automated machine intended to construct framework structures in space, possibly for future space **platform**s. Although such a device has yet to be used operationally in space, a development model has been demonstrated. The machine supplies three continuously extruded longitudinal beams formed from flat spools of aluminium alloy, which are spot-welded into a triangular-section structure using pre-formed cross-braces. The length of the beam is theoretically limited only by the length of the spools and the number of cross-braces the machine can hold.
[See also **space station.**]

BEAM CENTRE

The point on the Earth's surface where the signal power in the **beam** of a satellite **antenna** is at a maximum—its **peak.**
[See also **beamwidth, footprint, boresight.**]

BEAM EDGE

Often used in an inexact manner but, most usefully, the beam edge is the locus of points on the Earth's surface where the signal power in the **beam** of a satellite **antenna** is half its **peak** value.
[See also **beamwidth, beam centre, footprint.**]

BEAM-HOPPING

The practice of rapidly switching a satellite **downlink** transmission between different **beam**s in a multiple-beam **antenna**. See **satellite switching**.

BEAM WAVEGUIDE

A type of **waveguide** used in **earth stations** between the **high-power amplifier (HPA)** and the **antenna** to reduce RF losses [figure B1]. Instead of propagating the normal 'waveguide modes', which involves a current distribution in the accurately fashioned walls of the guide itself, the beam waveguide simply provides a 'weatherproof tunnel' for a shaped **beam** of RF radiation. A narrow beam is formed by a **feedhorn** and is reflected 'optically' through any bends in the guide until it reaches the antenna, where a **subreflector** intercepts it and reflects it onto the main reflector. The main reflector forms the **uplink** beam in the usual way.

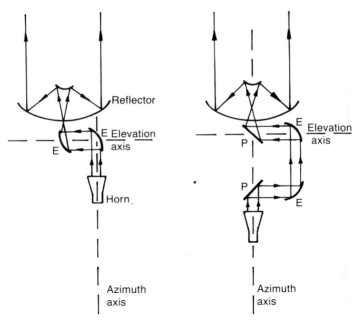

Figure B1 Beam waveguide configurations. P = plane reflector; E = elliptical reflector (or offset **paraboloid**).

BEAMWIDTH

The width of the **beam** of radiation shaped by a communications **antenna** (units: degrees). Unless an antenna is designed to be **omnidirectional**, it concentrates the radiated signal into a beam with a certain **directivity**. Naturally, the beam widens as it moves away from the antenna, so that by the time the signal reaches its destination it has spread out considerably. The measure of spreading is represented by the 'beamwidth' of the antenna, important in that it indicates whether two ends of a link will be able to communicate: they must lie within the beamwidths of each other's antennas to do so. Moreover, the converse tends to reduce the possibility of **interference** between two systems which do not wish to communicate.

The most useful way to define the concept of beamwidth involves a numerical statement of how the radiated power decreases away from the beam centre or **boresight**: the 'half-power beamwidth' (HPBW) is the width of the beam to the points where the radiated power is half its **peak** value. The concept is applicable to all communications antennas, not just the commonly seen parabolic dishes. See **antenna**.

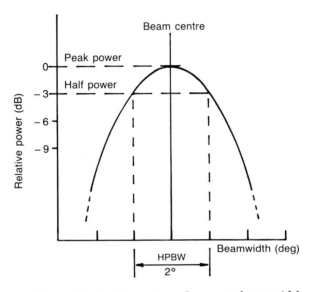

Figure B2 An illustration of antenna **beamwidth**.

The power decreases from the centre of the beam (the peak) towards the edge, as shown in figure B2: at the half-power point (-3 dB) the beamwidth can be determined. For circular parabolic antennas the beamwidth is related to antenna diameter and **wavelength** by the following expression:

$$\text{HPBW} \simeq 72\,\lambda\,/D$$

where λ is the wavelength and D is the **earth station** antenna diameter (both in m). The constant 72 in the above expression assumes a circular aperture with parabolic illumination; if the illumination is uniform across the aperture, the constant is 58.4. The same expression can be used for linear apertures, but the constants are 66 for parabolic illumination and 51 for uniform illumination.

The projection on the Earth of a satellite antenna's HPBW delineates the **beam area**. The HPBW is a benchmark in the **radiation pattern** of an antenna since, to a certain extent, it allows the spacing of satellites and earth stations to be arranged so that they do not interfere with other services. The true criterion by which antennas are judged is, however, the **sidelobe** performance.

[See also **footprint**.]

BEANIE CAP

A conical attachment on a **swing-arm** (known as the **gaseous oxygen arm**) on the fixed **service structure** at the American **Space Shuttle**'s launch pad which extracts vented oxygen vapour from the nose of the **external tank (ET)** prior to launch. The release of the **cryogenic** vapour from the **liquid oxygen** tank would result in the freezing of the air around the nose of the ET, and subsequent launch vibrations could cause the resultant falling ice to damage the tiles of the **orbiter**'s **thermal protection system**. The beanie cap, which seals the nose of the ET with an inflatable collar, removes the vented gas before it can have this effect. The swing-arm is retracted about two minutes before launch.

[See also **vent and relief valve**.]

BEARING AND POWER TRANSFER ASSEMBLY

See **BAPTA**.

BEGINNING OF LIFE (BOL)

The beginning of a **spacecraft**'s operational **lifetime**. The term is also used to specify the magnitude of a parameter at the beginning of a spacecraft's lifetime, as opposed to at **end of life (EOL)** when it may be different (e.g. the output power from a **solar array** [see **degradation**], the mass of **station-keeping** propellant, etc).

BENDS

See **decompression sickness**.

BER

See **bit error rate**.

BERYLLIUM (Be)

A low-density metal increasingly used (when alloyed with other metals) in the aerospace industry. It can tolerate high temperatures and has high specific stiffness, high specific heat capacity and good thermal conductivity. It is, however, susceptible to surface damage from machining which can lead to brittle fracture. This, coupled with its toxicity, has made it more difficult to use than, say, **aluminium** or **titanium**.

Typical applications: components, mechanisms, casings and mirrors. [See also **materials**.]

BINDER

A substance which binds the components of a composite solid propellant together. See **solid propellant**.

BIPROPELLANT

One of two chemical **propellant**s (**fuel** or **oxidiser**) which are stored separately and combined for combustion to take place (e.g. **(liquid hydrogen + liquid oxygen; UDMH + nitrogen tetroxide)**. Bipropellants are predominantly **liquid propellant**s, but a liquid oxidiser and a solid fuel are used together in a 'hybrid' propulsion system [see **hybrid rocket**].

[See also **bipropellant thruster, combined (bipropellant) propulsion system, liquid apogee engine (LAE)**.]

BIPROPELLANT PROPULSION SYSTEM

See **combined (bipropellant) propulsion system**.

BIPROPELLANT THRUSTER

A propulsive device utilising two chemical **propellant**s—an **oxidiser** and a **fuel**; a type of **reaction control thruster**. Common bipropellants are monomethyl hydrazine (MMH: $CH_3NH.NH_2$) as the fuel and nitrogen tetroxide (N_2O_4) as the oxidiser. Since they are **hypergolic** (they ignite spontaneously when mixed), the thruster is simplified by the lack of an **ignition system**.

[See also **hydrazine thruster, hiphet thruster, orbital manoeuvring system (OMS)**.]

BIT

The smallest unit of information in a computer or communications system which can have one of two values (1 and 0) in binary notation. A contraction of b(inary dig)it. Usually seen with a prefix (for instance, kbits, Mbits, Gbits).

[See also **bit rate, baud, byte**.]

BIT ERROR RATE

A concept used to express the accuracy of digital **demodulation** or decoding; a measure of how truthfully the received signal represents the signal originally transmitted.

BIT RATE

A measure of the rate at which information passes through a computer or communications system: a data transfer rate. [Units: bits s^{-1}, kbits s^{-1}, Mbits s^{-1}, etc.]

Examples of the bit rates required for various applications show that, logically enough, bit rate is related to the complexity of the information carried. For example, speech via mobile radio and reasonable quality **telephony** can be provided at rates as low as 2.4 and 9.6 kbits s^{-1} respectively; **slow-scan TV** and fast **facsimile** require 64 kbits s^{-1}; hifi sound 250 kbits s^{-1}; a **teletext** picture about 1 Mbit s^{-1}; a 625-line colour TV signal 68 Mbits s^{-1}; and a high-definition TV (HDTV) system about 144 Mbits s^{-1}.

[See also **bit**, **baud**, **bit error rate**.]

BLAST-OFF

See **lift-off**.

BLOCKAGE

See **aperture blockage**.

BLOCKHOUSE

See **launch control centre (LCC)**.

BLOWDOWN SYSTEM

A closed **liquid propellant** delivery system which uses a fixed mass of **pressurant** through its period of operation; a system in which the pressure derived from the pressurant decreases as the tank is emptied (i.e. the **ullage pressure** decreases as the **ullage** volume increases).

Some **combined (bipropellant) propulsion system**s used in geostationary satellites operate with a renewable supply of pressurant for the apogee engine firing and then switch to 'blowdown mode' for the **reaction control thruster** firings made throughout the satellite's **lifetime**. Other delivery systems use one or other method all the time.

BLUE STREAK

A British intermediate range ballistic missile (IRBM) developed in the

1950s (**propellant**: **liquid oxygen/kerosene**). When the United Kingdom's independent nuclear deterrent was disbanded in 1960, Blue Streak was proposed as the first **stage** of a European three-stage launcher called **Europa**, which was then developed under the auspices of the European Launcher Development Organisation (**ELDO**). Following multiple failures of the upper two stages supplied by France and West Germany, Britain withdrew from ELDO but continued to supply Blue Streak boosters. By the early 1970s Europa had still not succeeded in orbiting a payload, so the programme was cancelled and ELDO was disbanded. ELDO and **ESRO** (European Space Research Organisation) were amalgamated to form the European Space Agency (**ESA**), which subsequently initiated the **Ariane** programme and delegated its management to **CNES** (the French national space agency).

BNSC

An acronym for British National Space Centre, the UK space agency formed in November 1985 to direct civil space activities and develop future policy. The BNSC brought together the civil activities of a number of established bodies: the Department of Trade and Industry (DTI), the Ministry of Defence (MOD), the Science and Engineering Research Council (SERC) and the Natural Environment Research Council (NERC).

BOIL-OFF

The evaporation of **cryogenic propellant**, usually from a **launch vehicle**, into the atmosphere. The evaporant is released through a 'boil-off' valve prior to final **pressurisation** and **launch** [see **vent and relief valve**].

BOILERPLATE

A term applied to a version of a **spacecraft** used only for test purposes; an engineering test vehicle (e.g. in the early American manned spaceflight programmes, **launch vehicle**s were fitted with (unmanned) 'boilerplate **capsule**s' which were the same size, shape and weight as the real capsules but contained few internal fittings or equipment).

BOL

An acronym for **beginning of life**.

BOLOMETER

A type of resistance thermometer, used particularly for the detection of radiant energy at infrared **wavelength**s: on Earth in infrared **telescope**s and in space as **earth sensor**s (as part of an **attitude control** system).

BOLT-CUTTER

See **pyrotechnic cable-cutter**.

BOLTZMANN'S CONSTANT

A constant named after Ludwig Edward Boltzmann (1844–1906), an Austrian physicist, which relates kinetic energy and temperature, and is in effect a 'conversion factor'.

Just as there is an absolute reference for temperature—'**absolute zero**' ($-273.15°C$ or 0 K)—there is an absolute reference for '**noise**' in a communications system. The lowest level of **background noise** is **thermal noise**, due to the motion of molecules which constitute any physical object. As the absolute temperature on the kelvin scale reaches zero, the molecular motion ceases and brings the thermal noise down to its zero, absolute reference. This is represented by Boltzmann's constant (-228.6 dBW Hz^{-1} K^{-1}), a constant of proportionality between the molecular kinetic energy and temperature, such that KE = $^3/_2\ kT$. The units of the constant describe, in terms of power (dBW), a level of noise contained in every 1 Hz of **bandwidth** for every 1 K of temperature.

Boltzmann's constant (in SI units, 1.381×10^{-23} J K^{-1}) is the ratio of the universal gas constant (8.315 J K^{-1} mol^{-1}) to Avogadro's number (6.022×10^{23} mol^{-1}), and should not be confused with the Stefan–Boltzmann constant ($\sigma = 5.67 \times 10^{-8}$ W m^{-2} K^{-4}), which is used to define the energy radiated by a black body.

BOOM

As part of a spacecraft, a pole or spar designed to hold an instrument or detector required to operate some distance from the main body of the vehicle. For example, mounting a **magnetometer** on a boom ensures that the spacecraft's contribution to the measured magnetic field is kept to a minimum.

BOOST PHASE

(i) The portion of a **launch vehicle**'s flight when **thrust** is being exerted by the **propulsion system**; a period of powered flight, as opposed to **coasting**.

(ii) The powered portion of a ballistic vehicle's **flight**.

[See also **ballistic trajectory**.]

BOOSTER

(i) A propulsive component of a **launch vehicle**, usually a **strap-on** booster added to the **core vehicle** to augment its **thrust**. Sometimes called a 'booster stage'.

(ii) A colloquial term for 'launch vehicle'.

[See also **solid rocket booster, rocket engine, rocket motor, retro-rocket**.]

BORESIGHT

The main axis of an **antenna**: a distinction is made between the **radio-frequency (RF)** boresight [see **peak**] and the geometrical boresight (physical axis) of the antenna, the angle between the two being known as the '**squint**' angle.
[See also **beamwidth**.]

BORESIGHTING

A colloquial term for the pointing of an **antenna**. See **boresight**.

BOX

Jargon: an area in the sky, as viewed from an **earth station**, which bounds the excursions of a **satellite** in **geostationary orbit** from its mean position. The satellite is not precisely stationary with respect to the earth station, mainly due to the gravitational forces acting upon it: the satellite drifts both above and below the equatorial plane within the limits of its **orbital inclination.** It executes a figure-of-eight pattern in the sky with the cross-over in the eight at its mean position as viewed from the ground [see figure B3]. This pattern is repeated daily. A common specification requires the satellite to remain within a box subtending an angle of $+0.1°$ from the Earth's centre, which equates to about 125 km (77.5 miles) at geostationary height.
[See also **orbital control**.]

BPSK (BINARY PHASE SHIFT KEYING)

A method for modulating the **radio-frequency carrier** in digital **satellite communications** links; a type of **phase modulation**. In BPSK or 'two-phase modulation', where sample carrier waveforms are 180° out of phase, one **bit** of data is encoded on the carrier for each phase change, allowing the representation of a '1' or '0'.
[See also **modulation**, QPSK **(quadrature phase shift keying)**.]

BREADBOARD

A lab-bench assembly of components or circuits of a preliminary design intended to prove the feasibility of the design, without regard to the final packaging or **configuration**. The term is derived from the practice of early electronics engineers, who mounted their experimental circuits on a wooden board usually used for cutting bread.

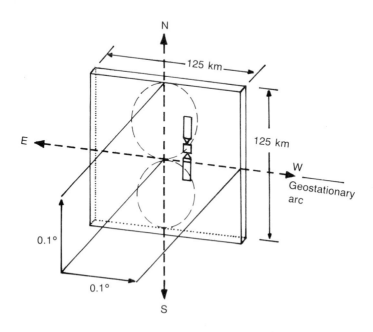

Figure B3 The station-keeping 'box' of a satellite in **geostationary orbit**, showing the tolerances on its **orbital position**.

BROADBAND

A modifier indicating the broad **bandwidth** of the quantity under consideration (for example, broadband **signal**, **broadband noise**, etc). Essentially the same as **wideband**.

BROADBAND NOISE

Noise which is spread across the band of a communications system; noise which occupies the full **bandwidth**. Also called wideband noise. In general, the wider the band the greater the noise, so in most cases the minimum bandwidth capable of accommodating the **signal** is used. [See also **frequency band**.]

BROADCASTING

The transmission of a **radio-frequency** signal carrying radio or television programmes (for instance) over a wide area of reception. The transmission medium is a specified part of the **electromagnetic spectrum**, there being no physical connection between transmitting and receiving **antennas**, in contrast to **cable** or **optical fibre** networks. [See also **narrowcasting**, **cable TV**, **direct broadcasting by satellite**.]

BROADCASTING-SATELLITE SERVICE (BSS)

A **satellite communications** system in which signals transmitted by the satellite are intended for direct reception by the general public. In the formal **ITU** definition the term 'direct reception' encompasses both individual and community reception.
[See also **direct broadcasting by satellite (DBS)**.]

BSS

See **broadcasting-satellite service**.

BUILT-IN HOLD

A pre-planned halt in the **countdown** to a **launch** of one of the more complex vehicles, included to allow time for any procedures which have fallen behind to 'catch up', and for final checks (e.g. on the weather) to be made.
[See also **hold**, **recycle countdown**.]

Figure B4 The Soviet space shuttle **Buran** on its Antonov An-225 carrier aircraft. [Mark Williamson]

BURAN

The name given to the first **flight model** of the Soviet **space shuttle** orbiter fleet [see figure B4]. Buran, which means 'Snowstorm', was launched on its first unmanned test flight on 15 November 1988 by an **Energia** launch vehicle.

BURN

Jargon: the operation of a **rocket engine** to alter the **attitude** or **trajectory** of a **spacecraft** (an 'engine burn', 'de-orbit burn', etc). Tends to be used only for propulsion *in vacuo* (i.e. not for **launch vehicles**). [See also **apogee motor firing, mid-course correction, plane change, de-orbit**.]

BURN-OUT

The moment at which a propulsive device ceases to provide a **thrust**, usually because its **propellant** supply has been exhausted; the cessation of propellant **combustion** (also called 'engine cut-off'). Used particularly for **launch vehicle** stages (for example 'third **stage** burn-out').
[See also **throttling, engine re-start (capability)**.]

BURST

A very-short-duration transmission in a communications system. For instance, in **time division multiple access (TDMA)** the basic repetitive unit is the 'frame' which consists of a number of bursts, one from each **earth station** connected to the system. A typical system, using **QPSK** at a burst rate of 60 Mbits s^{-1}, may have a frame duration of 15 ms, which is equivalent to 900 000 bits/frame. The first burst in each frame is transmitted by a controlling earth station and followed by bursts from the other stations in sequence. Although individual bursts are separated by 'guard times' to avoid overlapping at the receiving station, it is essential that all earth stations are time synchronised.

BUS

(i) Jargon for the part of a spacecraft that 'carries' or supports the **payload**. For example, on a **communications satellite**, the **service module** is known as the 'bus'; it may also be referred to as 'the **platform**'.

(ii) Jargon for the power supply and distribution system of a spacecraft. See **power bus, bus voltage**.

BUS VOLTAGE

The voltage rating of a spacecraft's main electrical power distribution circuit, or **power bus**. The range of bus voltages acceptable to spacecraft equipment is dependent on whether the power subsystem is 'regulated' or 'unregulated' (see **regulated power supply**).

The great majority of spacecraft power supplies are of the direct current type (see **DC power**). DC bus voltages have increased gradually over the years, more or less in line with the general rise in available DC power. A common standard for satellites manufactured in the USA is the 28 V system; the main European standard is the 50 V power-bus.

BYTE

A unit of information in a computer or communications system typically six or eight bits in length. The smallest unit of information that can be addressed or given a memory location. Normally two or four bytes are assembled to form a 'word', one of the basic units of data. [See also **bit**, **bit rate**, **baud**.]

C

C-BAND

4—8 GHz. See **frequency bands**.

CABIN

A general term for the habitable volume of a **manned spacecraft**, used particularly for small vehicles (as in 'spacecraft cabin'); for larger vehicles the alternative term, '**flight deck**', is more common.
[See also **capsule**.]

CABIN PRESSURE

The pressure of the atmosphere maintained within a **manned spacecraft**. Early American spacecraft (**Mercury**, **Gemini** and **Apollo**) were supplied with a pure oxygen atmosphere at a pressure of about 40 kPa (5.5 psi). The American space shuttle **orbiter** uses an oxygen/nitrogen atmosphere at 101.4 kPa (14.7 psi; 'sea-level pressure'). Soviet spacecraft (**Vostok**, **Voskhod**, **Soyuz**, **Salyut** and **Mir**) have always used an oxygen/nitrogen atmosphere at sea-level pressure [but see **decompression sickness**].
[See also **pre-breathe, decompression sickness, environmental control and life support system (ECLSS)**.]

CABLE

A medium for the transmission of television, data, etc, based on lengths of **coaxial cable** that provide a physical link between transmitting and receiving locations. The 'cable' may, increasingly in future, be a fibre-optic cable.
[See also **cable TV, optical fibre**.]

CABLE TV

A television service transmitted to the subscriber by **coaxial cable**; sometimes abbreviated to **CATV**. Since the **bandwidth** of the cable is, in practice, greater than that available for broadcast television, a cable system usually offers an increased number of TV channels to the recipient.
[See also **SMATV**, **broadcasting**.]

CABLE-CUTTER

See **pyrotechnic cable-cutter**.

CAPACITY

In **telecommunications,** a measure of the ability of a circuit, **transponder**, etc, to carry information or communications **traffic** (telephone calls, TV channels, data, etc). The capacity of a **communications satellite**, for example, is a function of many factors, such as the available **bandwidth** of its transponders, the **modulation** scheme, the satellite transmission power and the size of the **earth station** antenna.

CAPCOM

An abbreviation for 'capsule communicator'; a person assigned the task of relaying messages between '**mission control**' and a **manned spacecraft** (coined when spacecraft were called **capsule**s).

CAPE CANAVERAL

See **Kennedy Space Center**.

CAPE KENNEDY

See **Kennedy Space Center**.

CAPSULE

A name given to early **manned spacecraft** (or those carrying animals), due largely to their small size and their 'enclosing and protecting' nature (e.g. '**Mercury** capsule', '**Apollo** capsule', etc). Now generally used with reference to the proposed 'escape capsule' for the US/International **Space Station**.
[See also **cabin, flight deck**.]

CAPSULE COMMUNICATOR

See **CAPCOM**.

CARBON–CARBON COMPOSITE

A material composed of a carbon matrix reinforced by threads or fibres of carbon. The carbon matrix can be produced by 'hydrocarbon cracking' or by pyrolising a resin to leave a carbon residue into which the fibres are embedded. This class of composites has been developed for their high-temperature resistance: in excess of 1400 °C in an inert (non-oxidising) environment.

Typical applications: rocket motor **nozzles, thermal protection systems.** [See also **composites, carbon composite, ceramic–matrix composite, metal–matrix composite, materials.**]

CARBON COMPOSITE

A material composed of threads or fibres of the element carbon set in a 'matrix' of thermo-setting polymers such as the polyesters, phenols and epoxies.

Carbon fibre, one of the many materials used to reinforce polymer matrices, was developed in 1963 at the Royal Aircraft Establishment (RAE), Farnborough, UK. In the modern factory process carbon fibres are produced, for example, from specially treated fibres of polyacrylonitrile (PAN), a carbon-based polymer, known as an acrylic precursor. The controlled oxidation of the PAN precursor converts the acrylic fibres into a heat-resistant polymer which retains its textile character, making it possible to wind, weave or knit the final product. The level of oxidation can be adjusted to determine the precise properties of the resultant carbon fibre, dependent upon its eventual application.

Many similar terms will be found in the literature appertaining to composite materials used in spacecraft: carbon fibre, carbon composite, graphite composite, graphite epoxy and carbon-fibre-reinforced plastic (CFRP). In the main these descriptions refer to the same class of materials, since they are all based on the chemical element 'carbon'. However, in a more accurate sense, *carbon* products tend to be based on the relatively untreated 'amorphous carbon', whereas *graphite*-based products have been subjected to high-temperature and/or high-pressure processing to produce a more ordered structure, with the properties of higher electrical and thermal conductivity. The differences between particular products may also include the degree to which the precursor has been oxidised, the proportions of carbon and graphite, the constitution of the matrix, etc.

Typical applications: spacecraft **face-skin**s, **antenna** reflector surfaces and structural components.
[See also **materials, carbon–carbon composite.**]

CARBON DIOXIDE ABSORBER

A device in a manned spacecraft's **life support system** which filters

carbon dioxide from the air and recirculates it for further use. Colloquially called a 'CO$_2$ scrubber'. See **environmental control and life support system (ECLSS)**.
[See also **carbon dioxide narcosis**.]

CARBON DIOXIDE NARCOSIS

Unconsciousness induced by an excess of carbon dioxide in the atmosphere (i.e. more than about 2 kPa (0.3 psi) partial pressure, the maximum recommended concentration being about 0.7 kPa/0.1 psi). The normal sea-level partial pressure of carbon dioxide is about 0.03 kPa (0.004 psi), compared with the total air pressure of 101.4 kPa (14.7 psi).
[See also **carbon dioxide absorber, environmental control and life support system (ECLSS)**.]

CARBON-FIBRE-REINFORCED PLASTIC (CFRP)

A carbon-composite material used widely in spacecraft structures, mainly due to its characteristics of low density and high strength. See **carbon composite.**
[See also **materials**.]

CARGO BAY

See **payload bay**.

CARGUS

See **Sanger**.

CARRIER

In **telecommunications**, an electromagnetic wave of fixed **amplitude, phase** and **frequency** which, when modulated by a **signal**, is used to convey information through a radio transmission system. The frequency of the carrier is the frequency at which the **transmitter** operates, so the carrier conveys all the information held by the **baseband** signals and the **subcarrier**s which are modulated onto it.
[See also **carrier-to-noise ratio, modulation**.]

CARRIER GENERATOR

See **local oscillator**.

CARRIER-TO-INTERFERENCE RATIO

See **protection ratio**.

CARRIER-TO-NOISE-POWER-DENSITY RATIO (C/N_0)

The ratio of the power level of a signal **carrier** wave to the **noise power density (NPD)**, represented by the expressions:

$$\frac{C}{N_0} = \frac{P_C}{(P_N/B)} = \frac{P_C}{kT} = \frac{C}{kT}$$

where P_C is the carrier power, P_N is the **noise power**, B is the **bandwidth**, k is **Boltzmann's constant** and T is the **absolute temperature**. See **noise power density.**

The difference between C/N (**carrier-to-noise ratio**) and C/N_0 is that C/N_0 considers only the **noise** in 1 Hz of bandwidth (noise/Hz), whereas C/N takes account of the total noise in the full bandwidth.

CARRIER-TO-NOISE RATIO

The ratio of the power level of a signal **carrier** wave to the power level of the **noise** at the same time and place in a radio transmission system. Abbreviated to C/N or, less frequently, CNR, this quantity refers to a **signal** at RF or IF as opposed to **baseband**.
[See also **noise power, signal-to-noise ratio, carrier-to-noise-power-density ratio, radio frequency, intermediate frequency.**]

CARRIER WAVE

See **carrier.**

CASSEGRAIN REFLECTOR

A style of dual reflector **antenna** using a convex hyperboloidal **subreflector** and paraboloidal main reflector [see figure N1]. The **feedhorn**, subreflector and main reflector are coaxial in the standard design, but the feed can be offset (see **offset feed**); see figure A2. This geometry was first described for optical astronomical telescopes in 1672 by Professor N Cassegrain of the Collège de Chartres to the Paris Académie des Sciences, although the first practical example was not constructed until the mid 18th century, by James Short.
[See also **Gregorian reflector, Newtonian telescope, Schmidt telescope, Coudé focus, space telescope.**]

CATV

Originally an acronym for community antenna television, but now also used as an abbreviation for cable television.
[See also **cable TV, SMATV.**]

CCIR

A French acronym for International Radio Consultative Committee, a body within the International Telecommunications Union (**ITU**), which considers and issues recommendations on technical, operational and tariff matters relating to **radio-frequency** use.

CCITT

A French acronym for International Telegraph and Telephone Consultative Committee, a body within the International Telecommunications Union (**ITU**), which considers and issues recommendations on technical, operational and tariff matters relating to international telegraph and telephone services.

CDMA

See **code division multiple access**.

CELESTIAL BODY

Any astronomical body (e.g. stars, planets, moons, etc). Historically, the Earth has not been included, since it has not been thought of as a 'heavenly body'—it is, after all, the centre of the 'celestial sphere'. However, from a point in outer space, the Earth is just another celestial body.
[See also **planetary body**.]

CENTAUR

A rocket **stage** used with the Atlas **launch vehicle**. See **Atlas**.

CENTAUR G/G-PRIME

Two versions of a type of **upper stage** derived from the Centaur upper stage used on the **Atlas** launch vehicle. Originally intended for use with the American **Space Shuttle**, but as a result of new safety requirements instigated following the destruction of the Space Shuttle **Challenger** its use was confined to expendable launch vehicles (ELVs). **Propellant: liquid oxygen/liquid hydrogen. Payload** capability: Centaur G, over 4500 kg to **geostationary orbit** (GEO); Centaur G-prime, nearly 6000 kg to GEO.
[See also **payload assist module (PAM), inertial upper stage (IUS)**.]

CENTRE OF GRAVITY

The point through which the resultant of the gravitational forces on a body always acts. Commonly abbreviated to 'c-of-g'. Where gravitational forces are very low (i.e. in the 'weightless condition'), the term **centre of mass** is substituted.
[See also **gravity, low gravity, microgravity**.]

CENTRE OF MASS

The point through which the resultant of any applied forces on a body always acts. Commonly abbreviated to 'c-of-m'. For example, for best effect the thrust vector of a spacecraft's **thruster** should act through the c-of-m; when subject to an **attitude control** manoeuvre, a spacecraft rotates about an axis through its c-of-m.
[See also **centre of gravity**.]

CENTRE SPATIALE GUYANAIS (CSG)

See **Guiana Space Centre**.

CENTRIFUGAL FORCE

A force on a body moving in a circle or along a curved path which acts directly away from the centre of curvature (literally 'centre-fleeing'). The opposing force is called 'centripetal', from the Latin for 'seeking the centre'; since the acceleration of a body in circular motion is defined mathematically as 'acting towards the centre of rotation', it is termed 'centripetal acceleration'.

In physics, centrifugal force is known as a 'fictitious force', because its effects occur only as a reaction to the opposing centripetal force (a 'true force'). Take, as an example, a spacecraft travelling in accordance with Newton's first law: it will continue to move in a straight line until a force opposes that motion. If a planet were to 'suddenly appear' to one side of the spacecraft, its gravitational attraction would pull the spacecraft into an orbit around it. The true force here is obviously that due to gravity, which acts towards the centre of the planet. Of course, the spacecraft does not move in the direction of this force, because it also has an orbital velocity. It is this which creates the apparent outward, 'balancing' force which stops it moving towards the planet—this is the fictitious centrifugal force. If the planet's gravity were to be instantaneously removed, the spacecraft would continue at a tangent to its previous motion, in a straight line, thereby proving that the radial, centrifugal force existed only as a result of the centripetal (gravitational) force.

In a **centrifuge**, or a spacecraft spun to induce 'artificial gravity', the centrifugal force is again a reaction against the centripetal acceleration produced by circular motion. The bodies involved would fly off at a tangent were it not for a seat or floor pushing against them—it is this force which gives the impression of 'weight'.
[See also **weightlessness**.]

CENTRIFUGE

(i) A device which simulates the varying accelerations experienced in

spaceflight, particularly during **launch**. Although modern manned **launch vehicle**s such as the American **Space Shuttle** are designed not to exceed about '3g' during launch, early vehicles used for the launch of **manned spacecraft** greatly exceeded this value (e.g. '9g'): the centrifuge not only allowed **astronaut**s to experience the accelerations in a controlled manner, but also allowed them to develop a tolerance which would avoid a 'black-out' at a critical phase of the mission.

(ii) Any rotating device for separating liquids of differing density or liquids from solids by the action of **centrifugal force**.

CENTRIPETAL FORCE/ACCELERATION

See **centrifugal force**.

CEPT

A French acronym for European Conference of Postal and Telecommunications administrations, an international commission designed to strengthen relations between the European post and **telecommunications** administrations (PTTs).

CERAMIC—MATRIX COMPOSITE (CMC)

A material composed of a ceramic matrix reinforced by threads or fibres of a ceramic material (e.g. silicon carbide/silicon carbide). Fibre-reinforced CMCs offer high-temperature resistance without the inherent fragility of **ceramics**.
[See also **composites, carbon composite, metal—matrix composite, materials**.]

CERAMICS

A class of materials made from clay and similar substances; an inorganic compound or mixture requiring heat treatment to fuse it into a homogeneous mass. The majority of ceramics can withstand a much wider temperature range than most metals and are generally resistant to temperatures up to and above 2000 °C (in an oxidising environment): e.g. alumina, used in ablative shields, melts at 2054 °C.
[See also **ceramic—matrix composite, composites, materials, ablation**.]

CEU

See **control electronics unit**.

CFRP

See **carbon-fibre-reinforced plastic**.

CHALLENGER

(i) The name given to the second '**flight model**' of the American Space Shuttle **orbiter** (orbiter vehicle OV-099), named after a sailing ship which explored the Atlantic and Pacific Oceans in the 1870s.

Challenger was originally built as a 'structural test article' (STA-099), but after the completion of the design verification it was converted to an orbiter capable of **spaceflight**. It was first launched on 4 April 1983 (STS 31-B). On 28 January 1986 (mission 51-L, the 25th Space Shuttle mission) the failure of an 'O-ring' seal on one of the **solid rocket boosters** caused the vehicle to explode, with the consequent destruction of the spacecraft and the loss of its crew.

(ii) The name given to the **Apollo** 17 **lunar module**.

[See also **Space Shuttle, space transportation system (STS), Enterprise, Columbia, Discovery, Atlantis, Endeavour.**]

CHANNEL

A **band** of radio frequencies with a pre-determined upper and lower limit designed to carry a specific quantity or quality of information. Thus every channel has a **bandwidth** and, depending on how it is to be used, can carry a certain **bit rate**.
[See also **channel filter.**]

CHANNEL FILTER

A specific type of **band-pass filter** used to define the **bandwidth** of a communications **channel**. This **filter** is placed immediately before the transmit section of a **transponder** chain, which is largely responsible for amplifying the signal.

CHANNEL ISOLATION

A measure of separation between **radio-frequency** (RF) channels. When used qualitatively, it is expressed in terms of **frequency separation**: the channel centre-frequencies are separated from one another by a number of hertz (Hz). The band of frequencies separating the two nominal channel **bandwidth**s is known as a guard-band.

In a more quantitative sense, the isolation of RF channels is quoted in terms of the relative **powers** of their **carriers**: one channel is a number of dB above the other. This is otherwise known as **protection ratio**.
[See also **channel, isolation, guard-band.**]

CHANNEL SEPARATION

The practice of providing different **radio-frequency** channels to avoid **interference** between communications services which operate in the same geographical area or which use **satellite**s in the same **nominal orbital position**. This is especially important if the services operate on the same **polarisation**. Essentially the same as **frequency separation**. [See also **channel, channel isolation, interference, co-channel interference**.]

CHANNELISATION

The process of dividing up the **radio-frequency** spectrum into specifically defined portions of **bandwidth**, or **channel**s, for the efficient and non-interfering transmission of information.

CHARACTERISTIC VELOCITY

(i) A parameter which characterises the **combustion** performance of a **rocket engine** or **rocket motor**. Characteristic velocity (denoted by $C\star$ or 'C-star' and measured in m s^{-1}) is given by the expression

$$C\star = P_c A_t / \dot{m}$$

where P_c is the combustion chamber pressure (N m^{-2}), A_t is the nozzle throat area (m^2) and $\dot{m}$ is the propellant mass flow rate (kg s^{-1}). The higher the value of $C\star$, the higher the combustion efficiency.

If a rocket **nozzle** was to be cut off at the **throat**, the **exhaust velocity** would be equal to the characteristic velocity, thus allowing the effectiveness of the **propellant** to be demonstrated independent of the design of the exit cone.

(ii) The velocity a vehicle would acquire due to the **thrust** of its engines under ideal conditions (i.e. no **gravitational attraction**, no **drag**, etc).

CHARGE

See **propellant grain**.

CHASE-PLANE

An aircraft which follows and observes a spacecraft, particularly the American Space Shuttle **orbiter** during its descent and landing phase.

CHEMICAL PROPULSION

A form of **rocket** propulsion in which chemical propellants either undergo **combustion** or are decomposed to produce **thrust**.

[See also **propellant, rocket engine, rocket motor, hydrazine thruster, electric propulsion, nuclear propulsion**.]

CHILLDOWN

Jargon: the process of preparing or 'conditioning' **launch vehicle** and ground-based propellant systems for the loading of **cryogenic propellant**. When the equipment has reached a sufficiently low temperature, the propellant can be loaded without damaging the equipment. A 'slow-fill' typically precedes a nominal 'fast-fill' to further acclimatise the tanks.

CHUGGING

A form of **combustion** instability in a **rocket engine** which manifests itself as a low-frequency pulsing sound.

CIGARETTE COMBUSTION

See **propellant grain**.

CIRCULAR ORBIT

An **orbit** with a nominally constant radius, e.g. **geostationary orbit**, most **low Earth orbit**s.
[See also **elliptical orbit, parking orbit, drift orbit**.]

CIRCULAR POLARISATION

See **polarisation**.

CLARKE ORBIT

See **geostationary orbit**.

CLEANROOM

A room in which the standard of cleanliness, measured by the allowed number of particles of a certain size, is higher than the general indoor environment; an area typically used for the **integration** of **spacecraft**, **subsystem**s and other instruments whose performance could be detrimentally affected by 'dirt'.
[See also **white room**.]

CLOSED-ECOLOGY LIFE SUPPORT SYSTEM (CELSS)

See **life support system**.

CLOSURE PANEL

A section of a spacecraft's exterior which can be removed to allow access to components during the later stages of assembly. In practice, the fitting of these panels are amongst the last activities in the spacecraft **integration** process since they are used to 'close up' the spacecraft. For this reason they tend to be called closure panels rather than 'access panels', but the use of the terms is open to personal interpretation.

CMC

See **ceramic—matrix composite**.

CNES

An acronym for Centre National d'Etudes Spatiales, the French national space centre, a government organisation, founded in March 1962, responsible for the implementation of French space policy.

CO_2 SCRUBBER

A colloquial term for a **carbon dioxide absorber**.

COAST EARTH STATION

An **earth station** in the maritime **mobile-satellite service (MSS)**, usually located near the coast since the **satellite**s of the MSS **space segment** are located (in **geostationary orbit**) centrally over the world's oceans.
[See also **ship earth station**.]

COASTING

The motion of a **spacecraft** or **launch vehicle** due to its momentum or a force of **gravitational attraction** (i.e. when its **propulsion system** is inactive). The term is typically applied to launch vehicles during **stage separation** and to spacecraft travelling between one **planetary body** and another (NB if a spacecraft is in **orbit**, it is said to be 'orbiting', not coasting).
[See also **boost phase, ballistic trajectory**.]

COAX

Colloquial abbreviation for **coaxial cable**.

COAXIAL CABLE

An electrical **transmission line** with two coaxial conductors (on a common axis)—known as the 'inner' and the 'outer'—separated by an insulator (or 'dielectric' material). The inner is solid or stranded, often of copper-coated aluminium, and the outer is a flexible wire braid which allows bending to a certain degree. A domestic example is the cable used to carry the TV signal from an aerial to a television receiver.

Coaxial cable, often abbreviated to 'coax', is capable of carrying relatively low powers compared with a **waveguide**, and is commonly used, therefore, on the RF input side of a **travelling wave tube (TWT)**.

COAXIAL SWITCH

A device designed to isolate part of a circuit and divert a signal carried in **coaxial cable** (one of two types of commonly used electrical **transmission line**). Devices are either switched magnetically using solenoids, causing a mechanical deviation of the line, or electronically using diodes.
[See also **waveguide, waveguide switch**.]

CO-CHANNEL INTERFERENCE

Interference from a transmitter operating on or near the same channel **frequency**. Under normal conditions this type of interference is rare, since telecommunications and broadcasting services are allocated widely spaced **channel** frequencies if they are in close physical proximity. However, when meteorological conditions create an inversion in the lower atmosphere, signals can be 'guided' over larger than normal distances by a phenomenon known as 'ducting', where atmospheric layers act as a 'natural **waveguide**'.

CODE DIVISION MULTIPLE ACCESS (CDMA)

A coding method for information transmission between a potentially large number of users, whereby all the users occupy the total transponder **bandwidth** all the time (compare **time division multiple access** (TDMA) and **frequency division multiple access** (FDMA)).

In CDMA the signals are encoded so that they can only be received by authorised receivers. Primarily used for military communications.
[See also **spread spectrum, encryption**.]

CODEC

An electronic device used in **telecommunications** for **coding** and **decoding** digital transmissions (e.g. of **television**). A contraction of COder–DECoder.

CODING

See **encryption, scrambling, scrambler**.

COLD-GAS THRUSTER

A type of **reaction control thruster** which uses a gas (e.g. nitrogen, argon, freon, propane) stored at high pressure and fed, through a pressure reducer, to an expansion **nozzle** to produce **thrust**. It is termed 'cold' because there is no **combustion** or heating of the **propellant**.
[See also **hydrazine thruster, hiphet thruster, bipropellant thruster, ion thruster**.]

COLD SOAK

Exposure to an extended period of low temperatures (e.g. to reduce the temperature of a spacecraft component or **subsystem** to the local ambient, usually in an **environmental chamber** which simulates the conditions experienced in space).
[See also **heat soak**.]

COLD WELDING

A process whereby two metal surfaces 'diffuse into each other', due to a lack of lubricant or even an air gap between them. Although cold welding can occur on Earth when two surfaces are held in close contact under pressure, the space environment is especially conducive to the effect.

COLLECTOR

That component of a **travelling wave tube** (TWT) which collects and effectively 'disposes of' the **electron beam** after its interaction with the **RF wave** in the **slow-wave structure**. It typically consists of several 'stages' which collect electrons with different kinetic energies. To decelerate the electrons, collector stages are operated at negative potentials (i.e. voltages below 'ground'). This is known as a 'depressed collector'. The efficiency of the device can be enhanced by increasing the number of stages (although rarely beyond five) and tube efficiencies of over 50% are possible.

  Decelerating the electrons causes their kinetic energy to be converted to heat, which the collector must then remove from the TWT either by conduction to a baseplate or by direct radiation. In the case of a high-power space 'tube' the collector enclosure projects through the spacecraft panel to radiate directly into space.

CO-LOCATED

A term used to express a similarity in position of two or more **satellites** in **geostationary orbit**, such that satellites in the same **nominal orbital position** are said to be 'co-located'.

COLUMBIA

(i) The name given to the first '**flight model**' of the American Space Shuttle **orbiter** (orbiter vehicle OV-102), named after a sailing frigate launched in 1836, one of the first Navy ships to circumnavigate the globe. Columbia was completed and transported to **Kennedy Space Center** in March 1979 but, due to technical problems, was not launched until 12 April 1981. This was the first of four development flight tests (STS-1 to STS-4).

(ii) The name given to the **Apollo** 11 command module.

[See also **Space Shuttle**, **space transportation system (STS)**, **Enterprise, Challenger, Discovery, Atlantis, Endeavour**.]

COLUMBUS

The name given to the European Space Agency's contribution to the 'Freedom' International **Space Station** programme. The three **ESA** elements are the Columbus Attached Laboratory, the Columbus Free-Flying Laboratory and the Columbus Polar Platform.

COMBINED (BIPROPELLANT) PROPULSION SYSTEM

A satellite **propulsion system**, using liquid **bipropellant**s, which combines the functions of **apogee kick motor** (AKM) and **reaction control system** (RCS). Also known as a 'unified propulsion system' [figure C1].

The system's **liquid apogee engine** (LAE) is capable of several separate **burn**s, allowing the accurate transfer of spacecraft to **geostationary orbit** (GEO). By varying the timing and magnitude of the burns to adjust the periods of intermediate **transfer orbit**s, the final burn can be performed at the desired on-station longitude, injecting the satellite directly into its correct **orbital position**. This technique, known as **parasitic station acquisition**, can save **station-keeping** fuel and increase the satellite's **lifetime**. NB by contrast, a solid propellant AKM is capable of only a single burn [see **rocket motor**].

Another advantage of the LAE is that, once in GEO, it can be isolated from the rest of the system so that all propellant not used for the apogee burns can be made available to the **attitude and orbital control system**, which also increases the potential life.
[See also **blowdown system**.]

Figure C1 JC Sat, an HS393 series **spin-stabilised** communications satellite (without its **solar array**). One of two **apogee** engines of its **combined (bipropellant) propulsion system** is visible below the row of propellant tanks. Two station-keeping thrusters can also be seen. Photograph also shows the folded **antenna**, the **TWTA**s of the **communications payload** and the honeycomb construction of the spacecraft floors. [Hughes Aircraft Company]

COMBUSTION

Any process of burning (e.g. the combustion of a rocket **propellant**; the oxidation of a rocket **fuel**). The substances which result from the combustion process are usually called 'combustion products'.
[See also **combustion chamber**, **propellant grain**, **solid propellant**, **liquid propellant**, **oxidiser**.]

COMBUSTION CHAMBER

The part of a **rocket engine** in which **propellant**s are burned to produce 'combustion gases' which subsequently pass through an exhaust **nozzle** to provide the **thrust** for a **spacecraft** or **launch vehicle**.
(See also **exhaust velocity**, **regenerative cooling**.]

COMMAND

An instruction transmitted to a spacecraft from a controlling **earth station** or **mission control centre** to maintain or adjust the operational status of the spacecraft and its subsystems; an abbreviation for telecommand. See **telemetry, tracking and command**.

COMMAND AND SERVICE MODULE

See **Apollo**.

COMMAND MODULE

See **Apollo**.

COMMON CARRIER

A national organisation or private company granted the authority to provide public **telecommunications** services in its parent country; in Europe generally the **PTT**.

COMMONALITY

The practice of common design. Equipment or systems with a high degree of commonality are largely interchangeable and can operate together.

COMMUNICATION

The imparting or exchange of information. (Note: this is part of the verb 'to communicate'; the corresponding noun widely used in technical circles is '**communications**', which may also be used as an adjective, e.g. **communications satellite**, communications link, etc.)

COMMUNICATIONS

The collective term for the information transmitted via a telecommunications system (e.g. a **communications satellite**). In the widest sense of the definition, the type of information is immaterial: it may comprise, amongst others, **telephony, television, facsimile** (photographs or photocopies) or computer **data**.
[See also **communication**.]

COMMUNICATIONS CARRIER ASSEMBLY

See **snoopy cap**.

COMMUNICATIONS MODULE

See **payload module**.

COMMUNICATIONS PACKAGE

See **communications payload**.

COMMUNICATIONS PAYLOAD

On a spacecraft, an assembly of electronic and **microwave** equipment designed to provide **communications** between the spacecraft and a **planetary body**, or another spacecraft (see **inter-satellite link**) [see figure C2]; the revenue-earning part of a commercial **communications satellite**, comprising both **transponders** and **antennas**; also called a 'communications subsystem' [see **subsystem**]. Sometimes called a communications package, especially on a spacecraft other than a communications satellite where the 'package' exists to handle spacecraft data rather than provide a revenue-based communications service (e.g. on manned spacecraft and planetary probes).
[See also **input section, output section, receive chain, transmit chain, payload module, antenna module**.]

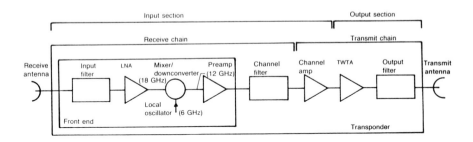

Figure C2 A satellite **communications payload** (LNA = **low-noise amplifier**; TWTA = **travelling wave tube amplifier**).

COMMUNICATIONS REPEATER

See **repeater**.

COMMUNICATIONS SATELLITE

A spacecraft designed to provide a communications service which may include, for example: telephony, television, telex, TV **distribution, facsimile, data communications** and **videoconference** services. Most communications satellites occupy a position in **geostationary orbit,** since there is no need for small **earth terminal**s to track a satellite in this orbit and a **constellation** of three spacecraft can provide complete coverage of the Earth, except for the high-latitude (polar) regions. However, the USSR has for many years used the high-inclination, elliptical **Molniya orbit** for communications satellites.

Communciations satellite is often abbreviated to 'comsat', not to be confused with the organisation **Comsat**.
[See also **communications payload, payload module, service module.**]

COMMUNICATIONS SUBSYSTEM

See **communications payload.**

COMMUNICATIONS SYSTEM

Any assembly of hardware and associated infrastructure designed to enable communications. A satellite communications system comprises a **space segment** and an **earth segment** (satellites and earth stations). [See also **communications payload, subsystem.**]

COMPANDING

A technique used in communications systems to improve the **signal-to-noise ratio** by compressing the **signal** at the **transmitter** and expanding it at the **receiver**—the word compand is a contraction of COMpress and exPAND.

COMPOSITE PROPELLANT

See **solid propellant.**

COMPOSITES

A class of **materials** in which fibres of a reinforcing material are set in a 'matrix' of another material. See **carbon composite, Kevlar composite, ceramic—matrix composite, metal—matrix composite.**

COMPUTER ENHANCEMENT

See **image intensification.**

COMSAT

(i) A colloquial abbreviation for **communications satellite.**

(ii) An abbreviation for the Communications Satellite Corporation, the United States' representative body to **Intelsat**. Comsat provides operational and technical services to Intelsat and operates its own domestic satellite system.

CONFIGURATION

The arrangement of a number of parts. In space technology, commonly used with reference to a **vehicle** of some description: e.g. a '**launch vehicle** configuration', the arrangement of its **stages**, **strap-on** boosters, etc; or a **manned spacecraft**, particularly the more complex, like a **space station**. Also used as a verb: e.g. 'the way the vehicle is configured'.

CONING

An instability experienced by a **launch vehicle** which combines motion in the **pitch** and **yaw** axes such that the vehicle's **roll axis** precesses about the direction of flight (like the **precession** of a **gyroscope**'s spin axis). The term is derived from the fact that the vehicle's longitudinal (roll) axis sweeps out the surface of a cone as it precesses.

CONSTANT CONDUCTANCE HEAT PIPE (CCHP)

See **heat pipe**.

CONSTANT-WEAR GARMENT

A one-piece zippered undergarment made of 'porous-knit' cotton, worn by **Apollo** astronauts beneath either a teflon in-flight coverall or a **spacesuit**.

CONSTELLATION

In **satellite communications**, a number of **satellite**s with approximately equal spacing around an **orbit** to provide maximum coverage of the Earth. The simplest constellation comprises three satellites in **geostationary orbit** with **orbital position**s about 120 degrees apart. More complex constellations have a greater number of satellites in lower altitude orbits: e.g. the Navstar Global Positioning System (GPS) is designed to comprise 18 **operational** satellites, equally divided between six 12 hour circular orbits (altitude 20 200 km, **inclination** 55°) [see figure C3].

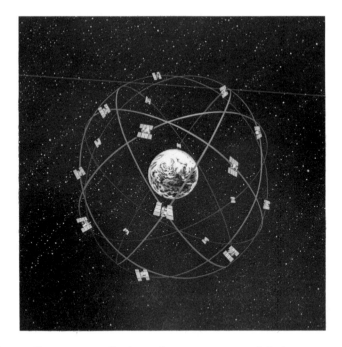

Figure C3 A **constellation** of Navstar GPS (global positioning system) satellites. [Rockwell International]

CONSUMABLE

Any item or supply which is consumed in an irreversible process, particularly in connection with manned spaceflight when used in a **life support system** (e.g. oxygen, water and food). It also includes **propellant**, some **coolant**s, etc.

CONTROL ELECTRONICS UNIT (CEU)

A device concerned with the overall control of a spacecraft—its central computer. It accepts **input**s from **sensor**s and directs **output**s to **actuator**s, as well as accepting **command**s from, and returning **telemetry** to, the Earth. With the increasing use of the microprocessor, spacecraft on-board control systems are becoming ever more complex. A typical modern system continually checks both input and output **data** and its own operation, so that it can correct errors and switch to a **redundant** unit in cases of failure.

CONTROL SYSTEM

Any mechanism or computer program which regulates the operation of

a device or assembly of devices (e.g. a **system** which controls the **attitude** of a **spacecraft** or the **flight path** of a **launch vehicle**).
[See also **attitude control, orbital control, control electronics unit (CEU), autopilot.**]

CONUS

An acronym for CONtiguous United States; the 48 states of the US mainland (i.e. excluding Alaska and Hawaii). The geographical arrangement of the USA has an impact on the design of many American **communications satellite**s in that **spot beam**s have to be provided for coverage of the outlying states.

COOLANT

Any fluid used as a heat-exchange medium to transfer heat from one part of a system to another (e.g. water in the cooling system of a **spacesuit**, or **liquid hydrogen** in a **rocket engine** [see **regenerative cooling**]).
[See also **cooling system, environmental control and life support system (ECLSS).**]

COOLING SYSTEM

Any device or structure which conducts or radiates heat in an effort to reduce the temperature of another device or structure. In spacecraft thermal design the term is usually applied to a device which refrigerates the component it protects. The devices most commonly in need of active cooling are those designed to detect heat itself (i.e. infrared **sensor**s) especially when they are used to detect very low levels of heat. The IR detectors used in infrared telescopes, for instance, must be cooled to liquid helium temperatures—a few degrees above **absolute zero**—so that the received signal is not completely drowned by the 'thermal noise' of the molecules in the detector itself. The term also includes non-refrigerating 'pumped fluid loops' used on **manned spacecraft**, space **platform**s, etc.

Open-cycle systems, which vent evaporated coolant to space, are not widely used because of their tendency to pollute the spacecraft's local environment and degrade optical surfaces and **solar cell**s. Closed-cycle systems resemble the domestic refrigerator in that they feature a coolant circulated through a closed loop, used as a heat-exchange medium. Spacecraft systems, however, are more compact and usually cool to much lower temperatures.
[See also **heat rejection, heat pipe, thermal control subsystem, cryogenics, bolometer, degradation.**]

COOLING SYSTEM (of a SPACESUIT)

Typically an undergarment worn next to the skin which is laced with capillaries through which water is pumped; otherwise known as a liquid cooling and ventilation garment (LCVG). See **spacesuit**.
[See also **environmental control and life support system (ECLSS)**.]

CO-POLAR(ISED)

Having the same **polarisation**. For example, two **RF wave**s with right-hand circular polarisation are said to be co-polarised. If one of the waves suffers **polarisation conversion**, it is said to have a co-polar component and a **cross-polar** component.

CORE VEHICLE

The part of a **launch vehicle** which is the same for all variants. Variants, which allow different payload sizes and masses to be launched, may be formed by the addition of **upper stage**s, **strap-on** boosters and/or different payload **shroud**s, etc.

CORNER-CUBE REFLECTOR

See **laser ranging retro-reflector**.

COSMIC RADIATION

Radiation from outside the Solar System comprising high-energy electrons and nucleons (chiefly protons); also called 'cosmic rays'.
[See also **single-event effect**.]

COSMODROME

See **Baikonur Cosmodrome**, **Northern Cosmodrome**.

COSMONAUT

The Soviet term for a person who has been trained for **spaceflight**; the title used by any person flying in a **spacecraft** operated by the USSR. The term is also used independently by the French.
[See also **astronaut**, **astronautics**, **cosmonautics**.]

COSMONAUTICS

The Soviet term for '**astronautics**'. Considered 'in translation', the Soviet term 'flight through space' is more general than the alternative 'flight to the stars' and more applicable to Man's present endeavours, but there is no sign of etymology taking precedence over common usage.

COSMOS

Name given to a series of Soviet spacecraft.

COSPAS-SARSAT

See **navigation satellite**.

COUDÉ FOCUS (TELESCOPE)

A type of astronomical reflecting telescope, based on the design of the **Cassegrain reflector**, in which light is reflected from a paraboloidal primary mirror onto a hyperboloidal secondary. The two designs differ in that the Coudé has an additional (plane) mirror below the secondary which reflects the light through an aperture in the side of the telescope to the eyepiece [see figure N1]. The second mirror moves to compensate for the motion of the telescope, ensuring that the exit-beam remains fixed. Large terrestrial telescopes can have more than one usable focus, such as the 200″ Palomar telescope which has Cassegrain, Coudé and prime foci.
[See also **space telescope**, **Newtonian telescope**, **Schmidt telescope**, **Gregorian reflector**.]

COUNTDOWN

The period, timed by a clock running in reverse or 'counting backwards', leading up to a **launch**; the act of 'counting down' to the moment of **lift-off**, particularly the verbal '5...4...3...2...1...' immediately prior to launch.

The exact timing of events around 'zero' varies from one **launch vehicle** to another: e.g. the first **stage** engines of the **Ariane** launcher are ignited at 'zero' and the vehicle is released at $T+3.4$ seconds, whereas the **Space Shuttle main engine**s are ignited at $T-6.6$ s and the vehicle is released at 'zero', when the **solid rocket booster**s are ignited. The time delay in both cases is that required for the **thrust** of the respective **liquid propellant** engines to build up to the level required to achieve lift-off.
[See also **countdown sequencer**, **hold-down arm**.]

COUNTDOWN SEQUENCER

A computerised system which controls the sequence of events preceding a **launch**. A typical system initiates specific actions at preprogrammed times, or prompts a human operator to do so, and indicates the status of all **launch vehicle** and **ground support equipment** to operators in a **launch control centre**. Due to the complexity of most modern launch vehicles, much of the process is automatic: only if the sequencer detects that a parameter is outside its pre-determined limits will it initiate a '**hold**' in the **countdown**.

COUPLED CAVITY

A type of **slow-wave structure**; a component of a **travelling wave tube** (TWT).

COVERAGE AREA

An area on the surface of the Earth, defined for the purposes of **satellite communications**, within which the **power flux density** (PFD) of the radiated satellite signal is sufficient to provide the desired quality of reception for a telecommunications service in the absence of **interference** [figure C4].

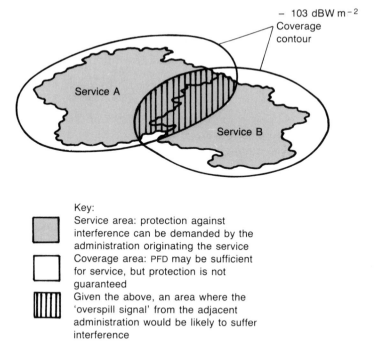

Figure C4 The relationship between the **coverage area** and **service area** of a satellite communications service as defined by the **ITU** (International Telecommunications Union).

The PFD may be sufficient for service but, in contrast to the **service area**, protection against interference is not guaranteed. One of the provisions of the **ITU** Radio Regulations states that the coverage area must be the smallest area which encompasses the service area, but since no satellite beam (**footprint**) can possibly adhere exactly to a country's political boundary, this necessarily leads to **overspill**. For instance, for the European DBS services defined by **WARC** 1977, the power flux density contour which delineates the coverage area is -103 dBW m^{-2}.
[See also **beam area, protection ratio**.]

COVER-GLASS

See **coverslip**.

COVERSLIP

A protective cover for a **solar cell**, bonded to its top surface; otherwise known as a cover-glass.
   Solar cells are susceptible to damage from the Sun's high-energy radiation and particle flux. The coverslip filters out short wavelength UV, to prevent darkening of the cover-glass adhesive, and IR wavelengths, which give rise to undesired heating of the cell. More generally, coverslips offer protection against micrometeoroids and, in **low Earth orbit**, the impact of residual atmospheric oxygen.
[See also **degradation**.]

COWLING

See **skirt**.

CRADLE

A U-shaped framework structure used to support a **spacecraft**, typically a **communications satellite** with a **perigee kick motor** (PKM) attached, in the **payload bay** of the American **Space Shuttle**. Also known as a 'cradle assembly' and sometimes referred to as **airborne support equipment** (ASE). The cradle also supports a retractable **sunshield** which protects the satellite before its release from the payload bay.
[See also **pallet**.]

CRAWLER

A vehicle used to transport the American **Space Shuttle**, mounted on its **mobile launch platform** (MLP), from the **vehicle assembly building**

(VAB) to the **launch pad** [see figure C5]; also known as the 'crawler transporter'. It is 40 m long, 35 m wide and 6.1 m high, and has an unladen weight of 2700 tonnes. Its payload, an MLP and an unfuelled Space Shuttle, amounts to almost 5000 tonnes. It has a dual-tracked drive unit at each corner and is powered by two 2750 horsepower diesel engines driving four 1000 kW generators which provide power to 16 traction motors: its maximum speeds are 3.2 km h^{-1} (unloaded) and 1.6 km h^{-1} (loaded). A levelling system keeps the Shuttle vertical as it negotiates the 5% gradient up to the launch pad, where it deposits the MLP and withdraws. Two Crawlers, based on the design of a stripping shovel used for surface coal-mining in Kentucky, USA in the 1960s, were built for the **Apollo** programme, to move assembled **Saturn V** launch vehicles from the VAB to the pad.

Figure C5 Space Shuttle and **mobile launch platform**, carried by a **crawler** transporter, climb the slope to the **launch pad**. Note that the crawler compensates for the incline to keep the launch platform level. [NASA]

CRAWLERWAY

A road at **Kennedy Space Center** which links the **vehicle assembly building** (VAB) with **launch pad**s 39A and 39B, so called because it is designed to be used by the '**Crawler**'. It comprises two 12 m wide lanes 15 m apart and measures about 5.6 km from the VAB to launch pad 39A and about 6.8 km to 39B.

CREW

The men and women aboard a **manned spacecraft**. See **astronaut**, **cosmonaut**, **mission specialist**.

CROSS-POLAR DISCRIMINATION (XPD)

See **discrimination**.

CROSS-POLAR(ISED)

Having the opposite **polarisation**. For example, an **RF wave** with right-hand circular polarisation and another with left-hand circular polarisation are said to be cross-polarised. If one of the waves suffers **polarisation conversion**, it is said to have a **co-polar** component and a cross-polar component. Since the waves were of opposite **polarity** to begin with, it is the co-polar components which may lead to **interference** between the two **signal**s.

CROSS-RANGE CAPABILITY

The ability of a **Space Shuttle** returning from orbit to **manoeuvre** to the left or right of the planned flight path. The American Space Shuttle **orbiter**, for example, has a cross-range capability of about 2000 km left or right of its nominal **glide path**. It was designed with delta wings to provide this, since it has no engines to allow a second approach attempt.
[See also **range**, **downrange**.]

CROSS-STRAPPING

The practice of interconnecting two or more **redundant** chains of equipment to allow switching between the chains in the event of a failure. Cross-strapping is common in **propulsion subsystem**s and in chains of electronic and **microwave** equipment (e.g. in **communications payload**s, **attitude and orbital control system**s, **tracking telemetry and command** systems, etc).
[See also **single-point failure (SPF)**.]

CRYOGEN

A substance at very low temperature or used to produce very low temperatures (e.g. liquid helium used to cool infrared detectors, etc).
[See also **cryogenic propellant**.]

CRYOGENIC PROPELLANT

A **liquid propellant** (**fuel** or **oxidiser**) formed by the liquefaction of a gas when cooled to very low ('cryogenic') temperatures.

The most widely used cryogenic fuel and oxidiser, respectively, are **liquid hydrogen** (LH$_2$) which boils at about -253 °C, and **liquid oxygen** ('LO$_2$' or 'LOX') which boils at about -183 °C. The LH$_2$/LOX combination is used for the **Space Shuttle main engine**s and the **Ariane** launch vehicle's third **stage**, for example.

Cryogenic propellants, in general, are difficult to handle, since they evaporate at environmental temperatures, and their use requires propellant tanks to be insulated, which adds to the structural weight of the vehicle. Hydrogen, in particular, has a very low density, requiring large tanks, which also adds to the weight. However, their relatively high **specific impulse** offsets the disadvantages.

A possible alternative to LOX is the potentially more potent oxidiser fluorine (F$_2$), but it is extremely corrosive and toxic, and therefore difficult to work with. The potential oxidiser ozone (O$_3$) produces the maximum energy release with any fuel, but is highly combustible and decomposes explosively in the presence of trace impurities.
[See also **liquid propellant, rocket engine, regenerative cooling**.]

CRYOGENICS

The branch of physics concerned with very low temperatures and the phenomena occurring at those temperatures.
[See also **cryogenic propellant**.]

CSG

A French acronym for Centre Spatiale Guyanais. See **Guiana Space Centre**.

CTPB

An acronym for the **solid propellant** carboxyl terminated polybutadiene. See **polybutadiene**.

D

DAMA

See **demand assignment multiple access**.

DANDE

An acronym for de-spin active nutation damping electronics; an
attitude control subsystem on a **spin-stabilised** spacecraft which
senses **nutation** using a **accelerometer** and uses the spacecraft's de-
spin motor to provide a cancellation force to reduce the nutation.

DATA

Information: in satellite **telecommunications**, generally used to mean
computerised information, a coded **bit**-stream, as opposed to other
traffic commonly handled by satellites such as **telephony** and
television. Often called **data communications**.

DATA COMMUNICATIONS

A type of **traffic** handled by many types of communications links
consisting generally of computerised or digital information, as opposed
to other traffic commonly handled by satellites such as **telephony** and
television. It is often abbreviated to 'datacoms'.

DATA RATE

The rate of information transmission through a communications system
measured in bits per second (bits s^{-1}) or any multiple (e.g. kbits s^{-1},
Mbits s^{-1}, Gbits s^{-1}).
[See also **bit**, **byte**, **data**, **data communications**.]

DATACOMS

Colloquial abbreviation for **data communications**.

DATUM

(i) A single piece or **bit** of information (the singular of **data**).

(ii) A known point or line from which other measurements or estimates can be made.

dB

See **decibel**.

dBc ('dB RELATIVE to CARRIER')

See **decibel**.

dBHz ('dB HERTZ')

See **decibel**.

dBi ('dB ISOTROPIC')

See **decibel**.

dBK ('dB KELVIN')

See **decibel**.

DBS

See **direct broadcasting by satellite**.

dBW ('dB WATTS')

See **decibel**.

DC POWER

A source of electrical power supplied with direct current. See **power**. The great majority of spacecraft power systems use DC, alternating current (AC) being used only for special applications.

DEAD-BAND

Jargon: the extent of any measured quantity between specified upper and lower bounds. For example, the extent of a geostationary satellite's apparent motion as viewed from the ground, otherwise known as its

station-keeping 'box'.
[See also **orbital control, guard-band, bandwidth**.]

DECAY

See **orbital decay**.

DECIBEL (dB)

A logarithmic ratio between two signal power levels used, amongst other things, to denote the **gain** of an **antenna, amplifier**, etc. (A gain of 3 dB is equivalent to a doubling in signal strength.) A numerical value (in dB) can be found using the following expression:

$$\text{No of dB} = 10 \log_{10} P_1/P_2$$

where P_1 and P_2 are the output and input powers respectively. A positive decibel value represents a gain; a negative value is a loss (e.g. 0.5 is equivalent to -3 dB, colloquially referred to as '3 dB down'). If two voltages are being compared instead of two powers, V_1 and V_2 replace P_1 and P_2 and '20' replaces '10' in the above expression.

   dBi is a measure of gain relative to an **isotropic** source; dBc denotes a measurement 'relative to the **carrier**'; dBW is the ratio of **power** output to a reference signal at 1 watt ($P_2 = 1$), expressed in decibels. In the same way that power is referenced to the watt in 'dBW' and to the milliwatt in 'dBm', **frequency** is referenced to the hertz in 'dBHz' and temperature to the kelvin in 'dBK'. Conversion from the numerical quantity in SI units to its equivalent in dB..., is performed simply by multiplying by $10 \log_{10}$ (e.g. as in the **bandwidth** of a **transponder**, 36 MHz = 75.56 dBHz; or the **noise temperature** of a **receiver**, 4 K = 6.02 dBK).
[See also **antenna gain** and **power flux density (PFD)**.]

DECLINATION

A measure of the angle north or south of the celestial equator, in the equatorial or geographical system of celestial coordinates (north is positive, south is negative). See also **hour-angle**: knowledge of a **satellite**'s declination and hour-angle allow an **earth station** to be aligned with it. An alternative 'horizon system' uses **altitude** (or **elevation**) and **azimuth**.

DECODING

See **decryption, descrambling, descrambler**.

DECOMPRESSION

The loss of atmospheric pressure from a **manned spacecraft** or spacesuit which may lead to **decompression sickness**, **hypoxia**, **anoxia** or **ebullism**.

DECOMPRESSION SICKNESS

A medical disorder caused by a sudden and substantial change in atmospheric pressure: characterised by severe pain, cramp and breathing difficulties. Otherwise known as 'dysbarism' or 'the bends'. The disorder, usually associated with deep-sea divers, develops when nitrogen absorbed in the blood and body tissues is suddenly released in bubbles as the ambient pressure on the body is reduced. In space, there is a continual danger of decompression of a **spacecraft** or spacesuit, especially considering the observed increase in **orbital debris**.

The problem was faced for the **Apollo–Soyuz test project** when spacecraft with different atmospheres were joined for the first time. The **Apollo** had an atmosphere of pure oxygen at about 37 kPa (5.4 psi), substantially lower than the average sea level pressure of 101.4 kPa (14.7 psi). The **Soyuz**, however, was pressurised with an oxygen/nitrogen mixture at about 70 kPa (10 psi). A direct transfer of **cosmonaut**s from the Soyuz to the Apollo would have given them 'the bends', so an intermediate **docking module** was designed to seal off the Apollo and equalise the pressure to that of the Soyuz before a meeting could take place. In addition, to simplify the operation, the Soviets reduced the pressure in the Soyuz to 65 kPa (9.5 psi).
[See also **hypoxia**, **ebullism**, **cabin pressure**.]

DECRYPTION

Decoding. Most military and some commercial radio transmissions are 'encrypted', to make them unintelligible to unauthorised receivers. They must then be 'decrypted'.

DE-EMPHASIS

In **telecommunications**, the restoration of a flat **baseband** response after **demodulation**; the opposite of **pre-emphasis**.

DEEP SPACE

The term is inexact, but typically refers to space outside the Earth–Moon system: that volume of **space** which is not 'in the vicinity of' Earth; not 'near-Earth space'. Thus **spacecraft** on interplanetary trajectories are usually tracked by a **deep space network** (DSN). In a less 'geocentric' definition, deep space could be defined as the volume of space outside the Solar System (e.g. 'the **Pioneer** spacecraft left the Solar System and headed into deep space').

DEEP SPACE NETWORK (DSN)

A network of **ground station**s used for **tracking** spacecraft in **deep space**, generally those on interplanetary trajectories. The DSN is operated for NASA by the Jet Propulsion Laboratory (JPL) in Pasadena, California, and has stations at Goldstone, California, Madrid, Spain and Canberra, Australia, sites which are also part of NASA's **space tracking and data network** (STDN).

DEGRADATION (of SPACECRAFT MATERIALS)

The deterioration of spacecraft **materials** under the influence of the space environment. For example, degradation of the **solar cell**, due to solar protons and electrons, energetic radiation and micrometeoroid impact is observed to decrease the power output of **solar array**s by 2 or 3% per year. For this reason it is common to quote two different values for spacecraft solar array output: one for beginning of life (BOL) and another for end of life (EOL). What this means in practice is that arrays have to be oversized by a considerable amount, based on BOL power, to ensure that the satellite will be fully operational at EOL.

Degradation of thermal control surfaces [see **surface coatings**] can lead to an increase in overall spacecraft temperature as the surfaces become less reflective, for example. Degradation of optical surfaces has an obvious effect on image quality.
[See also **coverslip**.]

DEGREE OF FREEDOM

An attribute of a body capable of linear or angular motion; an ability to move along an axis or rotate about it. A **spacecraft**, for example, has six degrees of freedom: three linear along each of three orthogonal axes (roll, pitch and yaw), and three rotational about those axes. A **launch vehicle** possesses similar degrees of freedom.
[See also **spacecraft axes**, **frame of reference**.]

DELTA

An American **launch vehicle** developed from the **Thor** rocket [see figure D1]. It uses **liquid propellant** in its first and second stages (**liquid oxygen/kerosene** and **nitrogen tetroxide/aerozine-50** respectively), and **solid propellant** in its third stage, which is typically a **payload assist module** (PAM). First-stage thrust is augmented by nine **strap-on** solid rocket boosters: six ignite at lift-off, burn out after about a minute and are jettisoned; the remaining three are then ignited.

The first variant of the Delta was launched in May 1960. Since then the many versions of the vehicle have launched approximately 200 scientific, weather and **communications satellites**. The Delta became a commercially operated launch vehicle in the late 1980s. The **payload** capability of the most powerful commercial variant (Delta 3920/PAM) is about 1270 kg into a **geostationary transfer orbit** (GTO) or about 730 kg to **geostationary orbit** (GEO).

[See also **Atlas**, **Titan**, **Scout**.]

Figure D1 Delta launch vehicle: note strap-on **solid rocket boosters**. [McDonnell Douglas]

DELTA MODULATION

A transmission method using a modulated radio-frequency **carrier** wave, whereby the **amplitude** of the original analogue input signal is sampled at discrete time intervals to create a representative digital translation of the signal. To this extent delta modulation is similar to **pulse code modulation** except that the sampling rate is typically 24 000 − 40 000 times a second (for good quality speech) compared with 8000 Hz for PCM. However, the major difference is that, having obtained the samples, a one-**bit** code is then used to transmit the change in input levels *between* samples. The 'delta' (△), used in the mathematic sense, refers to the change.

DELTA V (△V)

Mathematical terminology for a change in velocity. Typically used with reference to **launch vehicle**s and **spacecraft** (e.g. when quoting the velocity change required to transfer a spacecraft from one **orbit** to another).

DEMAND ASSIGNMENT MULTIPLE ACCESS (DAMA)

A coding method for information transmission between a number of users, whereby satellite **capacity** is assigned according to demand and not on a rigid 'pre-defined' basis. A spectrum-efficient method of dynamically allocating telephony **channel**s in a **transponder**.

DEMODULATION

The reverse process to **modulation** whereby a signal is recovered from a higher frequency modulated **carrier** wave. See **modulation**.

DEMODULATOR

A device whose function is to demodulate a signal from a **carrier** wave. See **modulation**.

DE-ORBIT

The procedure of leaving an **orbit** (e.g. 'to de-orbit a spacecraft', 'a de-orbit **burn**').
[See also **re-entry**, **graveyard orbit**.]

DEPLOY

(i) To move or extend a component, device or **assembly** away from the main body of a **spacecraft** (e.g. the 'deployment' of a **solar array**, **antenna**, **boom**, etc).

(ii) To place a **satellite** system '**on station**' (chiefly military).

(iii) To release a **payload** from the **payload bay** of the American Space Shuttle **orbiter**, or other similar **vehicle**.

[See also **deployable antenna**.]

DEPLOYABLE ANTENNA

A spacecraft **antenna** which 'unfolds' from a position of storage once the vehicle reaches its planned **orbit** or **trajectory** [figure D2].

Many spacecraft (e.g. **communications satellite**s) require large diameter antennas, so that they are able to produce narrow **spot beams**. Since they are generally too large to be mounted on the satellite's **antenna platform**, and also have to fit within the **shroud** of the **launch vehicle**, the only solution is to clamp them to the side of the satellite during launch and **deploy** them once in orbit [figure D3]. (The **solar array**s of the **three-axis-stabilised** satellite are stored and deployed in a similar fashion.)

[See also **unfurlable antenna**.]

Figure D2 Intelsat VI **deployable antenna**s during tests. Since the **antenna**s are not designed for the 1g environment, they are supported here by helium-filled balloons. [Hughes Aircraft Company]

Figure D3 The TDF-1 **DBS** satellite showing its **deployable antenna**s folded against the **feedhorn** tower. The folded **solar array** panels can also be seen. [Aerospatiale]

DEPTH OF DISCHARGE (DOD)

The amount by which a **battery** can be discharged without detrimentally affecting its future performance, measured as a percentage. For example, a DOD capability of 60% means that 60% of the maximum charge can be used. A battery's DOD capability has an effect on a spacecraft's **mass budget**, since if one battery has a capability twice that of another, a spacecraft will have to carry twice as many of the second battery to provide the same usable **power**.

DESCENT ENGINE

A spacecraft **rocket engine** used for a descent to the surface of a **planetary body**; the main engine in the spacecraft's 'descent stage' (e.g. the descent engine of the **Apollo** lunar module).
[See also **retro-rocket**, **ascent engine**.]

DESCRAMBLER

An electronic device used to restore radio transmissions, made unintelligible to unauthorised receivers by an electronic device known as a **scrambler**, to their original form.

DESCRAMBLING

A process by which radio transmissions, made unintelligible to unauthorised receivers by an electronic device known as a **scrambler**, are restored to their original form. Also used in telephone systems, etc.

DESIGN DRIVER

Jargon: any concept, practicality or physical law which fundamentally governs the design of a component or **system** (e.g. for most **spacecraft** the concept of 'minimum mass' is a design driver, due to the cost of launching unnecessary mass into space).

DESIGN LIFETIME

See **lifetime**.

DE-SPIN ACTIVE NUTATION DAMPING ELECTRONICS

See **DANDE**.

DE-SPUN PLATFORM

An **antenna platform** on a **spin-stabilised** spacecraft which remains stationary with respect to the Earth to maintain the **pointing** of the antenna **beams**. The shelf containing the communications **transponder** may also be de-spun, depending on the design. See figure D2 for an example.
[See also **BAPTA**.]

DIELECTRIC LENS

See **lens antenna**.

DIFFUSION MEMBER

A stiff structural component used to 'diffuse' concentrated **load**s over a wider area of structure (e.g. in the **thrust structure**s of launch vehicle **propulsion bay**s.)

DIPLEXER

A two-channel multiplexer. See **multiplexer**.

DIPOLE

A simple type of 'wire antenna' [see **antenna**], commonly called a 'half-wave dipole', which consists of a linear conductor approximately† half
†Resonance is typically attained at 0.49 wavelengths.

a **wavelength** long with the input connection in the middle (so-called 'centre-fed').

DIRECT BROADCASTING by SATELLITE (DBS)

A **satellite communications** system whereby **television**, and associated audio and data channels, are broadcast direct to the home, as opposed to via a landline or **microwave link**. It obviates the need for a large number of transmitters throughout the country with the inherent problems in providing full coverage in difficult terrain. The TV signal is transmitted from a single uplink **earth station** to the satellite, then retransmitted to the domestic antennas throughout the **service area**.

The home is equipped with a small parabolic **antenna** (typically between about 0.3 and 1 m in diameter), permanently pointed towards the satellite in **geostationary orbit**, coupled to indoor and outdoor electronics units. A **downconverter** attached to the antenna converts the high-frequency satellite signals to the lower frequencies accepted by the indoor electronics (**receiver**) linked to, or incorporated with, the TV set.

DBS satellites differ from other **communications satellite**s mainly in that they utilise high-powered **transmitter**s (usually with high-power **travelling wave tube**s) to provide high **equivalent isotropic radiated power** (EIRP). The desired simplicity of the receiving equipment led to the requirement for satellites with TWTAs of up to 260 W **RF power**, hence the term 'high-power DBS'. These high powers enable the small antennas and relatively inexpensive receiving installations to receive a good TV picture.

Although there are no hard-and-fast rules, satellite systems broadcasting with on-board powers of around 100 W tend to be called 'medium-power DBS', and those around 50 W 'low-power DBS'. If a satellite system not originally designed for DBS and using standard telecommunications TWTs provides a DBS service to small antennas, it is termed 'quasi-DBS'.

[See also **head-end unit, link budget**.]

DIRECT ORBIT

See **prograde orbit**.

DIRECTIVITY

The ratio of the radiation intensity from an **antenna** in a particular direction to that available from an **isotropic antenna**. A perfectly isotropic antenna can be said to have 'no **gain**' and 'no directivity'; any other antenna has definable values of both. In the theoretical case of a 'lossless antenna', the directivity and gain are the same.

DISCOVERY

The name given to the third **'flight model'** of the American Space Shuttle **orbiter** (orbiter vehicle OV-103), which was first launched on 30 August 1984 (STS 41-D). Discovery was named after two sailing ships: Henry Hudson's which attempted a search for a northwest passage between the Atlantic and Pacific Oceans in 1610–11, but instead discovered Hudson Bay; and Captain Cook's which discovered Hawaii and explored Alaska, etc.
[See also **Space Shuttle, space transportation system (STS), Enterprise, Columbia, Challenger, Atlantis, Endeavour.**]

DISCRIMINATION

A measure of the ability of a communications system or a component within that system to separate wanted from unwanted **signals**. Commonly used in the form 'cross-polar discrimination' (abbreviated to XPD) when comparing two **carrier** waves with opposite **polarisations**: the term relates to the degree to which the **feed** discriminates against the cross-polar component of the unwanted carrier.

DISH

A colloquial term for an antenna reflector, especially an **earth terminal**. See **antenna**.

DOCK

To link two **spacecraft** together in space. The term is used irrespective of the size of the spacecraft, and therefore includes devices such as the **manned manoeuvring unit** (MMU). In general there are two degrees of 'docking': the initial attachment, termed a 'soft dock'; and the establishment of a good mechanical contact which locks the spacecraft together, termed a 'hard dock'. For a **manned spacecraft** a hard dock is required before the crew can be safely transferred, since it forms an 'airtight seal'.

DOCKING MODULE

A part of a **spacecraft** containing one or more **docking ports**; alternatively termed a 'docking adapter' (e.g. the **Apollo–Soyuz test project** docking module, used to join together the two spacecraft, which had incompatible docking mechanisms; and the **Skylab** multiple docking adapter which allowed two **Apollo** spacecraft to dock with the

space station (although it was never done).
[See also **dock**, **docking probe**.]

DOCKING PORT

An access point on a **spacecraft** to which another spacecraft can be
attached or 'docked' for the transfer of **crew** or materials.
[See also **dock**, **docking probe**.]

DOCKING PROBE

The 'male half' of a **spacecraft** docking system, typically attached to the
'active vehicle' (the spacecraft which executes the docking
manoeuvres) [see figure A6]. The 'passive vehicle' contains a **drogue**
into which the probe is inserted. Typically, once the spacecraft are
'hard-docked' together, the probe and drogue can be removed to allow
access through a docking tunnel.
[See also **dock**, **docking port**.]

DOD

(i) An acronym for the United States Department of Defence (written
DOD or DoD), a major US user of **space technology** from **launch
vehicle**s to **satellite**s.

(ii) An acronym for **depth of discharge**.

DOPPLER SHIFT

A change in **frequency** due to the motion of a transmitter or receiver.

DOUBLE-BASE PROPELLANT

See **solid propellant**.

DOUBLE CONVERSION TRANSPONDER

See **upconverter**.

DOUBLE HOP

A signal route in a **communications satellite** system which includes a
pass through two **satellite**s: i.e. the signal is **uplink**ed to one satellite,
downlinked to an **earth station**, uplinked to a second satellite and
downlinked to a second earth station [see figure D4]. Since the signal
travel-time to **geostationary orbit** is about 0.12 s for each of the four
links, the whole process takes almost half a second. Since this makes
normal conversation difficult, 'double hopping' is not recommended

for two-way telephone calls and one of the satellite links is often replaced by a terrestrial link.
[See also **inter-satellite link**, **terrestrial tail**.]

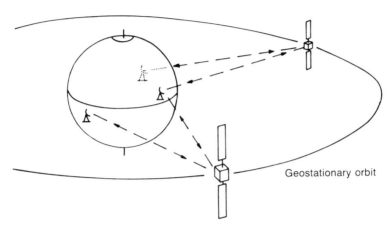

Geostationary orbit

Figure D4 A **double-hop** satellite communications link.

DOUBLER

A structural element which increases the thickness of a panel; local reinforcing for highly stressed areas.
[See also **thermal doubler**.]

DOWNCONVERTER

A device for reducing the **frequency** of a **signal**. It contains a circuit called a **mixer**, which 'mixes' the incoming signal with a **local oscillator** frequency. This process, known as the heterodyne process, or 'heterodyning', produces frequencies corresponding to the sum and the difference of the two original frequencies. The output of the downconverter is the difference signal; in a **transponder** with a downconverter and an upconverter this signal is otherwise known as the IF or **intermediate frequency**.
[See also **upconverter**.]

DOWNLINK

The communications path or link between a **satellite** and an **earth station**, 'from space to Earth'. Opposite: **uplink**.

DOWNRANGE

The distance between a **launch vehicle** and the **launch pad** (or the

ground station tracking the vehicle) measured on the surface of the Earth (e.g. 'the launch vehicle is 3 km downrange'); the distance measured along the projection of the vehicle's **flight path** on the ground, its **ground track**.
[See also **slant range, range**.]

DRAG

The retarding force acting on a body in motion through a fluid, particularly air. Drag acts in a direction parallel to the direction of motion, in opposition to the **thrust**.
[See also **atmospheric drag, drag compensation**.]

DRAG COMPENSATION

The practice of firing a spacecraft **thruster** to compensate for the effects of **atmospheric drag** experienced in **low Earth orbit**. More colloquially called 'drag make-up'.

DRAG MAKE-UP

A colloquial term for **drag compensation**.

DRIFT

(i) A gradual change in a spacecraft's **orbital parameters**, particularly the change in the longitudinal position of a spacecraft in **geostationary orbit**, due mainly to gravitational **perturbations** and the **solar wind**.

If this undesirable drift remained uncorrected, all geostationary satellites would tend to move away from their allocated **orbital positions** towards one or other of the stable equilibrium points. Drift is normally corrected by the periodic firing of the spacecraft's **reaction control thrusters**. See **equilibrium point, orbital control**.

(ii) The motion of a **satellite** in a **drift orbit**.

(iii) A lateral divergence of a **flight path** (due primarily to crosswinds).

(iv) A slow change in the **frequency** of a **transmitter** or other device containing a frequency source.

(v) The angular deviation of the spin axis of a **gyroscope** from its fixed reference in **inertial space**.

DRIFT ORBIT

An orbital path into which a satellite is injected by its **apogee kick motor** en route to its final position in **geostationary orbit**. The drift

orbit is an approximation to geostationary orbit in terms of height and circularity, but allows the satellite to **drift** towards its final **orbital position** without using its limited supply of **reaction control thruster** propellant.
[See also **equilibrium point, orbital control**.]

DROGUE

(i) Drogue parachute: a small parachute designed to pull a larger parachute from its container (e.g. for use with recoverable spacecraft and rocket boosters, etc).

(ii) Docking drogue: the 'female half' of a **spacecraft** docking system. See **docking probe**.

DRUM-STABILISED

See **spin-stabilised**.

DRY MASS

The mass of a **spacecraft** or **launch vehicle** without **propellant** or **pressurant**. In the case of a launch vehicle, where the concept of weight has a meaning, it is sometimes called 'dry weight'. Although **solid propellant** is, in effect, 'dry', it is treated in the same way as **liquid propellant** in this case.
[See also **mass budget**.]

DRY WEIGHT

The weight of a **launch vehicle** without **propellant** or **pressurant**. See **dry mass**.

DRYDEN FLIGHT RESEARCH CENTER

See **NASA**.

DSN

See **deep space network**.

DUAL-GRIDDED REFLECTOR

An **antenna** reflector designed for polarisation **frequency reuse** which comprises two reflector surfaces mounted one in front of the other. The front grid, which on the **uplink** reflects one polarity of radiation into its **feedhorn**, is transparent to the opposite polarity which is reflected from the rear grid into another horn. Angling the grids slightly with respect to each other allows a small separation between the horns while maintaining the same **footprint** for both polarities for the **downlink**. Dual-gridded antennas have been widely used on spin-stabilised communications satellites which generally have room for only one antenna on their top platform.

DUCTING

The entrapment of an electromagnetic wave between layers of the Earth's atmosphere or between the atmosphere and the Earth (effectively a 'natural **waveguide**'). The effect allows radio signals to be received at greater distances than they would be by 'line-of-sight'.

DUMMY STAGE

A launch vehicle **stage** which, as part of a test vehicle, contains no **propellant** and is therefore incapable of powered flight; a stage which simulates a launch vehicle **payload** for the purposes of a test flight. [See also **live stage**.]

DUTCH ROLL

A simultaneous rotation about the roll and yaw axes. See **spacecraft axes**.

DUTY CYCLE

The cyclic variation in the state of a device; the period from 'switch-on', through 'switch-off' to 'switch-on' again. For example, the duty cycle of a spacecraft **battery** is dependent on the **orbit**: in **geostationary orbit** a battery undergoes about 100 charge–discharge cycles per year, with varying periods [see **eclipse season**], whereas one in **low Earth orbit** typically undergoes around 6000 cycles with a 35 minute discharge period and a 55 minute charge period.

DYNAMIC PRESSURE

The pressure on a vehicle due to its motion through the atmosphere. The point in the **flight** of a **launch vehicle** at which it experiences the most severe aerodynamic forces is the point of 'maximum dynamic pressure', sometimes abbreviated to 'max-Q'. Since the **aerodynamic stress** is due to a combination of the ambient air pressure and the speed of the vehicle, most launch systems are designed to reduce **thrust** slightly as max-Q approaches and then increase it when the air pressure has dropped sufficiently (e.g. max-Q for the American **Space Shuttle** is reached at an altitude of about 10 200 m, about 60 seconds after **launch**).

[See also **thrust profile, throttling, propellant grain.**]

DYSBARISM

See **decompression sickness**.

E

EARLY BIRD

A colloquial name for Intelsat I, the first commercial geostationary **communications satellite**; launched in 1965 for the **Intelsat** organisation.

EARTH-LOCK

Jargon: 'locked onto the Earth'. A term used for **spacecraft** which require part of their structure (typically an **antenna platform**), to remain **pointing** towards the Earth. This is realised using infrared **earth sensor**s and an **attitude control** system.
[See also **spin-up** (sense iii).]

EARTH RESOURCES SATELLITE

A class of **remote-sensing** satellite concerned primarily with land-use. The term 'remote sensing' has tended to replace 'Earth resources'.

EARTH SEGMENT

The terrestrial part of a **satellite**-based communications system or any communications link with a **spacecraft**. Its major constituent is the **earth station**, but the earth segment may include several stations as well as the infrastructure which joins them to the user network. Earth segment is synonymous with **ground segment**.
[See also **space segment**].

EARTH SENSOR

A device used to establish a spacecraft's **attitude** relative to the Earth by detecting the limits of the Earth's disc at infrared wavelengths.

Sensing is either static or dynamic. The static **sensor** uses a common direction-sensing technique involving an arrangement of sensors grouped around the pointing axis so that they each receive the same illumination when the sensor is exactly aligned with the target source. An infrared image of the Earth is projected onto an array of thermopiles, which converts the incident IR radiation into an electric current. When the sensor is directly in line with the Earth the output from each of the thermopiles is the same; if the sensor is tilted with respect to the Earth the output changes accordingly. This type of differential output can be used to control a satellite's **reaction wheel**s, **reaction control thruster**s, etc. This is described as a 'null-seeking' technique. Since the unit has fixed optics, it can only be used in an orbit in which the Earth's apparent size remains constant (it is most prevalent in **geostationary orbit** from which the Earth subtends an angle of about 17.4°). This allows the Earth's disc to cover the sensing elements in identical proportions when the sensor axis is pointing towards the centre of the Earth, and to provide difference signals when it moves off axis. Since the temperature of the Earth's atmosphere varies with the time of day, a filter limits the incoming radiation to the carbon dioxide absorption band (around 15 μm) where the radiance is nearly constant.

The dynamic sensor operates by scanning across the Earth's disc and detecting the temperature difference between space and Earth. It is otherwise known as a 'horizon sensor' or 'horizon scanner'. When dynamic sensing is used on **three-axis-stabilised** spacecraft, usually those in **low Earth orbit**, the scanning motion is provided by a rotating mirror. **Spin-stabilised** spacecraft use two types of earth sensor, both of which make use of the satellite's inherent spin: the telescope type has a limited field of view and relatively high accuracy; the fan-beam type senses the Earth for a longer period but is less accurate.
[See also **sun sensor**.]

EARTH STATION

(i) An installation on the Earth containing the equipment necessary for communications with a **spacecraft**, i.e. chiefly an **antenna** system capable of transmission and reception, a **receive chain**, a **transmit chain** and the necessary interfaces with other terrestrial equipment [see figure E1]. An earth station is the major component of the **earth segment** (or ground segment) and is synonymous with ground station. Smaller versions are usually called earth terminals, a term sometimes used to include receive-only installations (e.g. **TVRO**).

(ii) The term 'earth station' is also used to refer to a facility comprising several earth-station antennas (e.g. Goonhilly earth station in Cornwall, England). 'Earth terminal' can be used in a similar way but usually refers to a smaller installation.

Figure E1 Earth station at Fucino, Italy. [B Paris]

EARTH TERMINAL

A small **earth station**.

EARTHSHINE

Thermal energy absorbed from the Sun and re-radiated by the Earth. See **albedo**.

EAST–WEST STATION KEEPING

See **orbital control**.

EASTERN TEST RANGE

See **Kennedy Space Center**.

EBU

An acronym for European Broadcasting Union, an association of **broadcasting** organisations whose members operate national broadcasting systems in Europe, North Africa and North America.

EBULLISM

The vaporisation of fluids from the human body under conditions of very low pressure (e.g. as experienced during a 'catastrophic' or explosive **decompression** of a **spacecraft** or **spacesuit**). When the ambient atmospheric pressure is less than the fluid vapour pressure of the body, about 6.3 kPa (0.9 psi or 47 mmHg), the fluids vaporise and bubble through the body's mucous membranes: the eyes, mouth and other orifices.
[See also **decompression sickness**, **hypoxia**.]

ECCENTRICITY

See **elliptical orbit**.

ECLIPSE

The partial or total obscuration of one celestial body by another, or of a spacecraft by another body [figure E2].

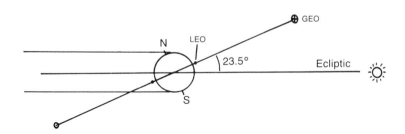

Figure E2 A satellite in an equatorial **low Earth orbit** goes into **eclipse** on every orbit. A satellite in **geostationary orbit** (shown to scale) is not eclipsed—this diagram shows the Earth at winter **solstice** [see figure E3].

Eclipse of the Sun by the Earth are particularly important for spacecraft which derive their **power** from **solar cells**. During eclipse most spacecraft draw power from batteries which are recharged once sunlight is available again. Eclipses also affect the **thermal energy balance** of a spacecraft by removing a source of heat.

Spacecraft in **low Earth orbit** enter the Earth's shadow for a similar time on every orbit: although it depends on their precise orbital period, they spend about 35 minutes of a 90-minute orbit in eclipse. For a spacecraft in **geostationary orbit**, eclipses occur on a less frequent but equally predictable basis—see **eclipse season**.

[See also **battery, thermal control subsystem**.]

ECLIPSE-PROTECTED

A **communciations satellite** service for which the satellite **transponder**s can remain powered during an **eclipse**, thereby providing a continuous service. The provision of 'eclipse protection' depends on the ability of the spacecraft's batteries to supply sufficient power for the **communications payload** and other subsystem equipment. High-powered DBS satellites are generally not eclipse protected, because the mass of the batteries required would be too great.

[See also **eclipse season**.]

ECLIPSE SEASON

The part of the year when a spacecraft in **geostationary orbit** (GEO) passes through the Earth's shadow and is unable to obtain **power** from the Sun via its **solar array**.

Since GEO is an **equatorial orbit** and the Earth's axis is tilted at 23.5° to the **ecliptic**, for most of the year the entire geostationary orbit is sunlit. However, as the Earth approaches the vernal (spring) equinox (21 March) and the autumnal equinox (23 September), part of the orbit moves into the shadow cast by the Earth and the satellite becomes eclipsed for a small part of its circuit [figure E3]. The angle of **inclination** of the equatorial plane relative to the Sun governs the duration of the eclipse, which is greatest when the plane exactly bisects the Sun. The 'eclipse season' begins about 22 days before the equinox and ends about 22 days after it, the eclipse duration increasing to a maximum at equinox when it peaks at about 72 minutes (penumbral eclipse is 71.8 minutes and that of the umbra 67.5 minutes).

In addition to the regular **eclipse** by the Earth, the Sun is eclipsed by the Moon, but on a much less regular basis.

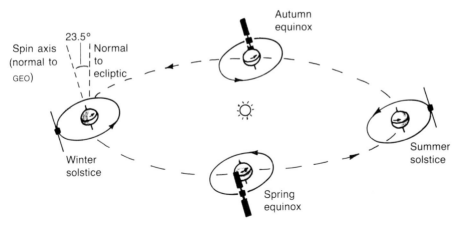

Figure E3 Eclipse geometry for a satellite in **geostationary orbit** showing the
eclipse seasons at the **equinox**es.

   Telecommunications satellites carry sufficient batteries to enable a
service to continue throughout the eclipse, which occurs around
midnight at the sub-satellite longitude. However, satellites using high-
power **transponders** (e.g. those for **direct broadcasting by satellite**
(DBS)) cannot carry the large number of batteries which would be
required to maintain a service. High-power DBS spacecraft are
therefore intended to be placed in **orbital positions** to the west of their
respective **service area**s so that they are eclipsed in the early morning,
when a 'closedown' is commercially less important.
[See also **solstice, solar array, sub-satellite point.**]

ECLIPTIC

In astronomy, the apparent annual path of the Sun relative to the stars
as seen from the Earth; the plane containing the Sun and the Earth is
known as the ecliptic plane.
   For historical reasons, the night sky visible from the Earth is known
as 'the celestial sphere'; the projection of the Earth's equator on this
sphere is called 'the celestial equator', and this and any similar
projection is known as a 'great circle'. The projection of the Earth's
orbital plane on the celestial sphere is a great circle—the ecliptic. Since
the Earth's axis is tilted at an angle of 23.5° to its orbital plane, the
ecliptic makes an angle of 23.5° with the celestial equator. It crosses the
celestial equator at two opposing points called the **equinox**es; the two
points on the ecliptic furthest from the equator are the **solstice**s.
[See also **eclipse, eclipse season.**]

ECLSS

See **environmental control and life support system**.

EDGE OF COVERAGE

Strictly the edge of the **coverage area** of a **satellite communications** service, but used more widely in the vaguer sense of 'the furthest point in all directions from the centre of a satellite **footprint** where a signal can still be received'. See **coverage area** and **service area**.

EDWARDS AIR FORCE BASE

The site of the NASA Ames Dryden Flight Research Facility [see **NASA**], where much of the USA's advanced flight research has been conducted. It includes the 65 square mile Rogers Dry Lake, an ideal natural surface on which to land high-performance research aircraft [e.g. see **X-1**, **X-15**] and the American Space Shuttle **orbiter** (e.g. during the 'Approach and Landing Tests' (ALT) [see **Enterprise**]).

EGRESS

An exit; the act of emerging. A word commonly used as a modifier (e.g. 'egress hatch', 'egress port'). The opposite of **ingress**.

EGSE

An acronym for electrical ground support equipment. See **ground support equipment**.

EHF (EXTRA HIGH FREQUENCY)

See **frequency bands**.

EHT

An acronym for electrothermal **hydrazine** thruster or electrically heated thruster. See **hiphet thruster**.

EIRP

See **equivalent isotropic radiated power**.

EJECTION VELOCITY

See **exhaust velocity**.

ELA

A French acronym for Ensemble de Lancement Ariane or '**Ariane** launch site'. The individual **launch complex**es at the **Guiana Space Centre** are thus known as ELA-1, ELA-2 and ELA-3.

ELDO

An acronym for European Launcher Development Organisation (also known by its French acronym CECLES), a body formed in April 1962 for the development of the **Europa** multi-stage **launch vehicle**. Founding countries were Australia, Belgium, France, Italy, Netherlands, the United Kingdom and West Germany. ELDO was disbanded in 1975 when it was amalgamated with **ESRO** (European Space Research Organisation) to form the European Space Agency (**ESA**).
[See also **Blue Streak**.]

ELECTRIC PROPULSION

A collective term for a form of rocket propulsion in which electrical energy is used to derive or augment the kinetic energy of the propulsive jet. There are three types of electric propulsion device:

(i) Electrothermal, in which electrical energy is used to heat a gaseous **propellant**, typically to increase the efficiency of a propulsive device (e.g. **hiphet thruster**);

(ii) Electrostatic, in which electrical energy is used to ionise a gaseous propellant and an electrostatic field accelerates the positive ions to produce **thrust** (e.g. ion engines or ion thrusters) [see **ion engine**];

(iii) Electromagnetic, in which an electromagnetic field is used to accelerate a neutral, gaseous **plasma** to produce thrust (e.g. plasma engines or plasma rockets).
[See also **chemical propulsion**, **nuclear propulsion**, **exhaust velocity**, **specific impulse**.]

ELECTRICALLY HEATED THRUSTER (EHT)

See **hiphet thruster**.

ELECTROMAGNETIC PROPULSION

See **electric propulsion**.

ELECTROMAGNETIC PULSE (EMP)

A short-duration burst of **electromagnetic radiation** which can damage electronic components that have not been 'hardened' against it. The term is used mainly in the context of the detonation of thermonuclear devices and their effect on electronics-based weapon systems, but can

be extended to more natural sources such as lightning.
[See also **hardening** (of satellites).]

ELECTROMAGNETIC RADIATION

Energy propagated through space or material media as an advancing disturbance in the intrinsic electric and magnetic fields, which oscillate in orthogonal planes mutually perpendicular to the direction of propagation.
[See **electromagnetic spectrum**.]

ELECTROMAGNETIC SPECTRUM

The entire range of **electromagnetic radiation**, from the shortest to the longest **wavelength** or the lowest to the highest **frequency** [figure E4]. In increasing wavelength (and decreasing frequency), sections of the EM spectrum are commonly labelled gamma rays, x rays, ultraviolet (UV), visible, infrared (IR), **microwave** and radio. All such radiations travel at the velocity of **light** in **free space**.
[See also **radio frequency (RF), frequency bands**.]

ELECTRON

An elementary particle which exists, in numbers equal to the atomic number, around the nucleus of every atom. The electron's rest mass† is 9.11×10^{-31} kg (1/1837 of proton's mass) and its charge is 1.60×10^{-19} C (symbol: e^- since the charge is negative).

The term 'electron' was suggested in 1891 by George Johnstone Stoney (1826–1911), an Irish physicist, as the name for a hypothetical particle believed to be a discrete unit of electricity. The electron was discovered experimentally at Cambridge University in 1897 by the English physicist Sir Joseph John Thomson (1856–1940).
[See also **electron beam, electron gun**.]

ELECTRON BEAM

A collimated stream of electrons produced by an **electron gun**. Used for example:

(i) as the source of energy for RF amplification in the **travelling wave tube**: the beam of electrons passes through the TWT from electron gun to **collector** along the axis of the helical **slow-wave structure**, where it interacts with the **RF wave**;

(ii) in the **ion engine** for the neutralisation of the positively charged ion beam.

†Its mass when 'at rest'.

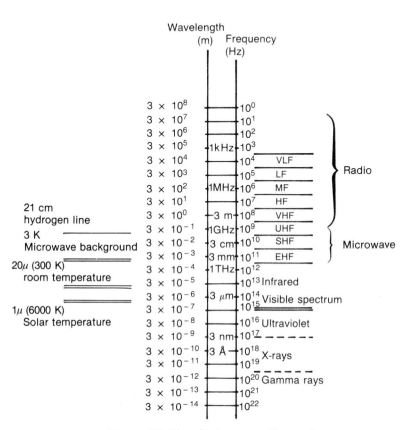

Figure E4 The electromagnetic spectrum.

ELECTRON BOMBARDMENT ION THRUSTER

See **ion engine**.

ELECTRON GUN

A device for producing a narrow beam of electrons from a heated cathode. Electrons liberated from the cathode are accelerated to high velocities by an electric field between the cathode and the anode. Although this definition covers equally the devices found in cathode ray oscilloscopes and domestic television receivers, the version used in space technology (e.g. within the **travelling wave tube** (TWT)) supplies an electron beam with a power density significantly higher than for the other two examples.

The electron gun in a TWT injects a narrow beam of electrons into a **slow-wave structure** where they are used to amplify an RF **signal**.

ELECTRONIC POWER CONDITIONER (EPC)

A device which supplies the operating voltages to the components of a **travelling wave tube** (TWT): the **electron gun**, **slow-wave structure** and **collector**. A TWT is, in effect, useless without an EPC and is never found in an operational condition in isolation.

[See also **travelling wave tube amplifier (TWTA)**.]

ELECTROSTATIC DISCHARGE (ESD)

An instantaneous loss of static electrical charge; a phenomenon which constitutes a potential hazard to **spacecraft** operations, generally to electronic devices but particularly to astronomical detectors and the like. Differential charging of the outside surface of the spacecraft can lead to a subsequent discharge which may, for instance, cause a communications system to switch off or severely disrupt an astronomical measurement.

The severity of the hazard to a communications **repeater** is represented by the Rosen scale, where 0 is 'no hazard', 1 is a 'nuisance' or outage of one second or less, 5 an outage of a few hours and 10 a catastrophe! The risk can be reduced to less than '1' by discouraging the build-up of charge: parts of the satellite should be conductive and grounded to the structure, particular attention being paid to shaded areas, which are not discharged by photoemission. Electronic circuits can also be protected against ESD by shielding and the addition of protective devices, such as filters and diode-like limiters, in the input and output lines.

ELECTROSTATIC PROPULSION

See **electric propulsion**.

ELECTROTHERMAL PROPULSION

See **electric propulsion**.

ELEVATION (ANGLE)

The angle above the horizon measured in the horizon system of celestial coordinates. The word **altitude** is synonymous with elevation in this context. See also **azimuth**: knowledge of a **satellite**'s azimuth and elevation angle allow an **earth station** to be aligned with it. An alternative 'equatorial' or 'geographical' system uses **declination** and **hour-angle**.

ELEVON

An aerodynamic control surface in the form of a hinged flap on the trailing edge of a delta-wing; a combined *elev*(ator and ailer)*on*. For example, the elevons on the American **Space Shuttle**, which are controlled by hydraulic power provided by an **auxiliary power unit (APU)**.

ELLIPTICAL ANTENNA

Any reflector **antenna** with an elliptical aperture. On a spacecraft, an elliptical antenna, in its simplest form, provides an elliptical **footprint**. However, the number, arrangement and detailed design of the **feedhorn**s used with a particular antenna make a wide variety of footprints possible with the same reflector.

The **major axis** of a simple elliptical antenna produces a narrower beamwidth than the **minor axis**, meaning that the resultant footprint is oriented orthogonally to the antenna reflector—see **beamwidth**.
[See also **boresight**.]

ELLIPTICAL ORBIT

A non-circular **orbit**; an orbit with a degree of eccentricity defined by a **major axis** and a **minor axis**. Due to irregularities in the gravitational attraction of a **planetary body**, all orbits have some eccentricity: a circular orbit is an approximation.
[See also **perturbations**.]

ELLIPTICAL POLARISATION

See **polarisation**.

ELV

See **expendable launch vehicle**.

EM

An acronym for (i) electromagnetic and (ii) engineering model, the version of a spacecraft constructed to verify that the electrical and RF performance meets the specification.

EMC

An acronym for electromagnetic compatibility.

EMI

An acronym for electromagnetic interference.

EMP

See **electromagnetic pulse**.

EMU

An acronym for **extra-vehicular mobility unit**.

ENCOUNTER

The moment in time or point in space at which a planetary exploration **spacecraft** on a fly-by **trajectory** is at its closest to the **planetary body** under investigation; the period over which the spacecraft is considered to be within close observing range of the body (the 'encounter period'). The term tends not to be used for spacecraft which enter an **orbit** about a planetary body.

ENCRYPTION

Coding. Most military and some commercial radio transmissions are 'encrypted' to make them unintelligible to unauthorised receivers. [See also **scrambling**, **descrambling**, **decryption**.]

END-BURNING

See **propellant grain**.

END EFFECTOR

A device fixed to the end of the American Space Shuttle's **remote manipulator system** (RMS) [see figure R3]: effectively the 'hand' on the mechanical arm. The standard end effector can 'grapple' a **payload**, keep it rigidly attached as long as required and then release it. For specific tasks, specialised end effectors can be fitted as necessary. [See also **grapple fixture**.]

END OF LIFE (EOL)

The end of a **spacecraft**'s operational **lifetime**. The term is also used to specify the magnitude of a parameter at the end of a spacecraft's lifetime, as opposed to at **beginning of life (BOL)** when it may be different (e.g. the output power from a **solar array** [see **degradation**], the mass of **station-keeping** propellant, etc).

END USER

The user of a communications system at the end of a link (e.g. the recipient of a **DBS** or **cable TV** service). A **cable** system operator, for example, could be classed as a user of a **satellite** since he rents a **transponder** for the distribution of his service. There is, however, a user of the system beyond the operator: the consumer—the end user.

ENDEAVOUR

The name given to the **Space Shuttle** orbiter built to replace **Challenger** and bring the American shuttle fleet back to four operational **orbiter**s. Otherwise known as orbiter vehicle OV-105, Endeavour was named after a sailing ship commanded by British Captain James Cook in the late 1700s.

ENERGIA

A Russian **launch vehicle** developed as a **heavy lift vehicle** (HLV) to carry, amongst other things, the Soviet **space shuttle**. It has three stages and four **liquid propellant** strap-on boosters and has a **payload** capability of over 90 tonnes to **low Earth orbit** (LEO). It was first launched, with an unmanned payload canister, in May 1987.
[See also **Buran, Proton**.]

ENERGY

The capacity of a body or system to do work, measured in joules (J).
[See also **power**.]

ENERGY DISPERSAL

A process which involves adding a low-frequency waveform to the **baseband** signal before **modulation**, to reduce the frequency-modulated signal's peak power per unit **bandwidth** and thus its **interference** potential. Particularly useful in avoiding the peaks of signal power which occur in a TV signal.
[See also **frequency modulation (FM)**.]

ENGINE

See **rocket engine**.

ENGINE BAY

The part of a **launch vehicle** or **spacecraft** which houses the **engine**s, typically incorporating a **thrust frame** and bordered by engine **cowling**s and other **fairing** components. Also called a propulsion bay.

ENGINE CUT-OFF

See **burn-out**.

ENGINE RE-START (CAPABILITY)

The ability to resume **combustion** in a **rocket engine**. It is only possible
to 're-light' engines using **liquid propellant**; once a **solid rocket motor**
has been ignited, it burns until all the propellant is consumed. Typically
only the uppermost **stage** of a **launch vehicle** has a re-start capability,
usually for injecting its **payload** into a **transfer orbit**. Spacecraft liquid
propulsion systems make continual use of this capability for **orbital
control**.
[See also **combined (bipropellant) propulsion system, liquid apogee
engine, parasitic station acquisition, hybrid rocket**.]

ENT (EQUIVALENT NOISE TEMPERATURE)

See **noise temperature**.

ENTERPRISE

The name given to the first-built American Space Shuttle **orbiter**
(orbiter vehicle OV-101), named after the *Starship Enterprise* from the
TV series *Star Trek*.

  Enterprise was used to determine the orbiter's aerodynamic
characteristics in a series of approach and landing tests (ALT)
conducted between February and November 1977 from a specially
modified Boeing 747, now known as the Shuttle Carrier Aircraft (SCA).
There were five 'captive flights' (with the orbiter fixed to the 747, inert
and unmanned), three manned captive flights and five 'release flights',
in which the orbiter was allowed to glide to a landing at **Edwards Air
Force Base**, California. In 1978 Enterprise was used for vibration tests,
and in 1979 transported to **Kennedy Space Center** for mating with an
external tank and **solid rocket booster**s for 'fit checks'. Enterprise was
never intended for **spaceflight** and is destined for the collection of the
Smithsonian Air and Space Museum, Washington, DC.
[See also **Space Shuttle, space transportation system (STS), Columbia,
Challenger, Discovery, Atlantis, Endeavour**.]

ENTRY INTERFACE

The point in a space vehicle's **trajectory** on its return to Earth which
defines the beginning of its **re-entry** (e.g. for the American **Space
Shuttle** it occurs about 30 minutes after the **de-orbit** 'burn' and about
two minutes prior to **S-band blackout**).
[See also **re-entry corridor**.]

ENVELOPE

A physical or conceptual boundary defining the limits of a parameter. For example:

(i) the dimensional constraint on the volume available for a **payload**— see **payload envelope**.

(ii) the operating limits of an **aerospace vehicle** (e.g. jargon: 'expanding the envelope').

ENVIRONMENTAL CHAMBER

A ground-based test facility which simulates some aspect(s) of the space or launch environment: namely temperature, pressure, humidity, noise and movement.
[See also **thermal-vacuum chamber, acoustic test chamber, vibration facility, anechoic chamber.**]

ENVIRONMENTAL CONTROL AND LIFE SUPPORT SYSTEM (ECLSS)

A system which maintains an air-conditioned environment (for crew and electronic equipment) and provides **life support** functions within the **cabin** of the American Space Shuttle **orbiter**. The ECLSS comprises atmospheric control and purification equipment and thermal control, water and waste management systems.

The nominal atmosphere in the orbiter comprises 80% nitrogen and 20% oxygen at a pressure of 101.4 kPa (14.7 psi). Recirculated air is passed through replaceable canisters containing a mixture of activated charcoal to remove odours, and lithium hydroxide to remove carbon dioxide. Cabin temperature can be regulated between 16 and 32°C; relative humidity between 35 and 55%. The air is passed through a **heat exchanger** and the excess heat is passed to a water **coolant** loop, from which it is transferred to a freon coolant loop and thence to the **radiator** panels mounted inside the **payload bay** doors. Water for food preparation, drinking and personal hygiene is produced by the **fuel cell**s, as a byproduct of energy generation. The 'waste collection system' collects and processes liquid and solid 'biowaste', washing water, and condensed water from the cabin heat exchanger.
[See also **portable life support system (PLSS), carbon dioxide absorber, spacesuit, cabin pressure.**]

EOL

An acronym for **end of life**.

EPC

See **electronic power conditioner**.

EQUATORIAL MOUNT

A structure for the support and guidance of an astronomical telescope which has its axis aligned with that of the Earth, with the result that the apparent motion of the stars can be 'cancelled out' by driving the telescope about one axis only (cf the **altitude-azimuth mount**). When used for **satellite communications** it is called a **polar mount**.

EQUATORIAL ORBIT

An **orbit** in the same plane as a planet's equator, e.g. **geostationary orbit**.
[See also **circular orbit, elliptical orbit**.]

EQUILIBRIUM POINT

One of four points on the **geostationary orbit** produced by the variation in the gravitational force around the Earth's equator, which is roughly elliptical in cross section [figure E5]. There are two 'stable equilibrium' points aligned with the minor axis and two 'unstable equilibrium' points aligned with the major axis. The approximate **orbital position**s of the stable points are 75°E and 105°W; those of the unstable points 165°E and 15°W (a 90 degree spacing around the orbit).

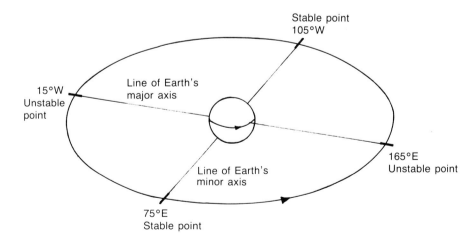

Figure E5 The **equilibrium points** on the **geostationary orbit**.

Although satellites stationed at the stable equilibrium points are less likely to **drift**, other **perturbations** of the orbit make **orbital control** necessary: correction of the drift around the orbit is known as east–west station keeping. For a satellite stationed at an arbitrary position between the equilibrium points, the drift mechanism operates in the following manner. If the effective component of the gravitational force is in the same direction as the satellite's motion (west to east), energy is added to the system and the orbital height increases. The satellite decelerates, its orbital period increases and it drifts back towards the equilibrium point on the minor axis. Passing the equilibrium point, the satellite experiences a force in opposition to its direction of motion, which reduces its energy. The orbital height decreases and the satellite accelerates, reducing its orbital period. It moves back towards the equilibrium point and the cycle continues. During its eastward drift, the satellite is up to 34 km below the nominal height of geostationary orbit, and during the westward drift it is above it to the same degree. The period of the oscillation is about 840 days, and the maximum rate is 0.4 degrees per day.

An **operational** satellite is not allowed to perform this oscillation, since it is required to remain within its station-keeping '**box**'. The necessary velocity increment (or change in velocity, $\triangle V$) required to correct the motion is a maximum of about 2 m s^{-1} per year depending on the **nominal orbital position**.
[See also **triaxiality, libration period**.]

EQUINOX

Either of two annual occasions when day and night are of equal length (vernal equinox, 21 March and autumnal equinox, 23 September).
[See also **solstice, eclipse, ecliptic, solar array**.]

EQUIPMENT BAY

See **instrument unit**.

EQUIVALENT ISOTROPIC RADIATED POWER (EIRP)

In **telecommunications**, the level of **transmitter** power, P, available at an **antenna** multiplied by the antenna **gain**, G (or added if working in decibels: EIRP(dBW) = P(dBW) + G(dB)). From the viewpoint of the **satellite**, it indicates the power (measured in dBW) of a particular combination of **transponder** and antenna; for an **earth station**, the antenna and the **high-power amplifier** (HPA), less any associated losses.

The 'benchmark' for **radiated power** is the **isotropic antenna**, which radiates equally in all directions and therefore has no gain. The EIRP gives the equivalent power which would have to be radiated from an isotropic antenna to provide the same **power flux density**. Increasing the **directivity** of the antenna, and thereby the gain, decreases the transmitter power required to provide an equivalent radiated power. (For example, suppose an EIRP of 50 dBW is needed: to provide this with an isotropic antenna (0 dB gain) would require a transmitter power of 100 000 W, which is about 400 times greater than is presently available from satellite transponders. However, if an antenna of readily attainable 30 dB gain is substituted, the required transmitter power is reduced to 20 dBW or 100 W. The example demonstrates the utility of the antenna.)

[See also **gain**, **decibel**.]

EQUIVALENT NOISE TEMPERATURE

An alternative term for **noise temperature**.

ESA

An acronym for European Space Agency, an organisation which provides for and promotes cooperation among member states in space research and technology and its applications. ESA was established in May 1975 by the member countries of **ESRO** (Belgium, Denmark, France, Italy, Netherlands, Spain, Sweden, Switzerland, the United Kingdom and West Germany) and by Ireland (Eire). It became a legal entity in October 1980 when its convertion was ratified, and gained two further members (Austria and Norway) in January 1987.

ESA combined the respective activities of ESRO and **ELDO**, in **satellite** and **launch vehicle** development, in a single organisation. Its main establishments are: headquarters (Paris); ESTEC (European Space Research and Technology Centre), Noordwijk in the Netherlands (design, development and testing of spacecraft); ESOC (European Space Operations Centre), Darmstadt in West Germany (satellite operations—**telemetry tracking and command**); and ESRIN (previously European Space Research Institute), Frascati in Italy (data collection 'Information Retrieval Service', and image processing 'Earthnet').

[See also **Guiana Space Centre**.]

ESCAPE TOWER

A structure mounted on top of a manned **expendable launch vehicle**, containing **rocket motor**s which can be used to pull the spacecraft **capsule** clear of the launch vehicle in case of emergency. Otherwise called a 'launch escape tower' or 'launch escape system'. Escape towers were used with **Mercury**, **Apollo** and **Soyuz** spacecraft; **Gemini**

and **Vostok** had ejection seats.
¹See also **Voskhod**.]

ESCAPE VELOCITY

The minimum velocity necessary for a body to escape from the **gravitational field** of a **celestial body**.

Three so-called 'space velocities' can be defined. The 'first space velocity' is that required for a body to become a **satellite**: at the Earth's surface, for example, it is about 7.9 km s⁻¹ (disregarding air resistance). The 'second space velocity' is that required to escape the gravitational pull of the celestial body: about 11.2 km s⁻¹ from the Earth's surface. The 'third space velocity' is that required to escape from the solar system: about 16.7 km s⁻¹ from Earth.

ESD

See **electrostatic discharge**.

ESOC

See **ESA**.

ESRIN

See **ESA**.

ESRO

An acronym for European Space Research Organisation, a body formed in June 1962 to promote, and provide the facilities for, collaboration among West European countries in space research and technology (also known by its French acronym CERS: Conseil Europeen de Recherches Spatiales). Founding countries were Belgium, Denmark, France, Italy, Netherlands, Spain, Sweden, Switzerland, the United Kingdom and West Germany. ESRO was disbanded in 1975 when it was amalgamated with **ELDO** (European Launcher Development Organisation) to form the European Space Agency (**ESA**).

ESTEC

See **ESA**.

ET

An acronym for the American Space Shuttle's **external tank**.

EUROPA

A European three-**stage** launch vehicle proposed in the 1960s and cancelled in the early 1970s. See **Blue Streak**.

EUROPEAN SPACE AGENCY

See **ESA**.

EUROSPACE

An industrial association established in September 1961 to encourage the development of space research and engineering in Europe.

EUROVISION

The system of regular television programme and news exchange in Europe set up in 1954 by the European Broadcasting Union (**EBU**). The first Eurovision-by-satellite tests performed using the Orbital Test Satellite (OTS) were followed by a regular service via the **Eutelsat** spacecraft.

EUTELSAT

An abbreviation for the European Telecommunications Satellite Organisation, a body established provisionally in June 1977, and on a permanent basis in 1982, by the members of CEPT (European Conference of Postal and Telecommunications administrations) to oversee and operate the **space segment** of European **telecommunications** embodied in the Eutelsat (formerly ECS) series of satellites.

EVA

See **extra-vehicular activity**.

EXHAUST

(i) The **combustion** products of a **rocket motor** or **rocket engine** ejected from an exhaust **nozzle** to produce **thrust**.

(ii) The waste products of a non-propulsive **combustion** process in a rocket engine, ejected from an 'exhaust pipe' (e.g. a **turbopump** turbine exhaust).
[See also **exhaust velocity, plume, plume impingement**.]

EXHAUST NOZZLE

See **nozzle**.

EXHAUST PLUME

See **plume**.

EXHAUST VELOCITY

The average axial velocity of the **combustion** gases at the exit of an exhaust **nozzle**. Also called 'ejection velocity'. Exhaust velocity (measured in m s^{-1}) is equivalent to specific impulse (N s kg^{-1}) [see **specific impulse**]. **Liquid propellant**s have higher exhaust velocities than **solid propellant**s.
[See also **characteristic velocity**.]

EXIT CONE

The divergent, bell-shaped part of a rocket exhaust nozzle which converts the thermal energy of the **combustion** gases to the kinetic energy of the exhaust **plume** and controls the expansion of the plume. See **nozzle**.
[See also **expansion ratio**.]

EXPANSION NOZZLE

See **nozzle**.

EXPANSION RATIO

(i) The ratio of the area of an exhaust **nozzle** exit to the **throat** area; otherwise called the 'area expansion ratio'.

(ii) The ratio of the pressure in a **combustion chamber** to that at the exit of an exhaust nozzle; otherwise called the 'pressure expansion ratio'. When the mean pressure in the exit area is equal to the ambient atmospheric pressure, the expansion ratio and the nozzle with that ratio are defined as 'rated'. A nozzle with an expansion ratio lower than rated is termed 'underexpanded' and one with a ratio greater than rated as 'overexpanded'.

[See also **flow separation**, **plug nozzle**.]

EXPENDABLE LAUNCH VEHICLE (ELV)

A **launch vehicle** which is only used once. Commonly abbreviated to ELV.
 The trajectories of ELVs tend to ensure that the discarded **stage**s and components fall into a sea or an ocean upon **re-entry**. Some upper stages 'burn up' in the atmosphere to an extent, but their remains are disposed of in the same manner.
[See also **reusable launch vehicle**.]

EXPERIMENTAL

Generally, 'apertaining to experiment'. When used in the context of space-based **hardware**, the term tends to be confined to scientific **payload**s and smaller scientific **spacecraft**. An advanced technology payload carried by a commercial **satellite** (e.g. a late 1980s **communications payload** operating in the 20 GHz **frequency** band) is typically termed 'a demonstration payload', rather than 'experimental', because of the lack of certain knowledge implied by the word 'experiment'.

EXTENSION MODULE

See **SYLDA**.

EXTERNAL TANK (ET)

The **propellant tank** which carries the **liquid hydrogen** and **liquid oxygen** propellants for the **Space Shuttle main engine**s (SSMEs), so-called because it is mounted externally to the Space Shuttle **orbiter**. The orbiter and two **solid rocket booster**s (SRBs) are attached to the tank in the launch configuration [figure S10]. The ET also acts as the structural 'backbone' of the Shuttle combination by absorbing the thrust **load**s from the SSMEs and SRBs.

Figure E6 Space Shuttle **external tank** at **Kennedy Space Center**, Florida. [NASA]

The ET is 47 m long and 8.4 m in diameter [figure E6] and contains two individual propellant tanks, one at the aft end containing about 1.5 million litres of liquid hydrogen (LH$_2$) and the other over 0.5 million litres of liquid oxygen (LO$_2$ or 'LOX') in the forward compartment. Subsequent layers of cork/epoxy and polyurethane-like foam insulation are sprayed onto the exterior surface of the ET. They insulate the propellants against the higher external temperatures, protect the tank from the effects of **aerodynamic heating** during launch and prevent the formation of ice prior to launch. Apart from increasing the vehicle's weight, the ice could vibrate loose during lift-off and damage the orbiter's **thermal protection system**. Tanks for the first two missions were painted white, a cosmetic coating which was eliminated for subsequent flights thus reducing the launch mass by about 270 kg.

The ET is the only expendable part of the reusable **space transportation system**, being discarded after its tanks are empty some 10 to 15 seconds after main engine cut-off (**MECO**). The philosophy of **staging** therefore remains, with parts of the launcher being jettisoned

Figure E7 A Space Shuttle **astronaut** conducting an **extra-vehicular activity** (EVA), in September 1985, to recover the Syncom IV-3 **communications satellite**, which was stranded in **low Earth orbit** when its **TT&C antenna** failed to deploy. [NASA].

throughout the launch phase to reduce the mass carried all the way to **orbit**. However, proposals have been made to carry ETs into **low Earth orbit**, where they can be converted for use as a type of orbiting **space station**.

[See also **cryogenic propellant, beanie cap, tumbling**.]

EXTRA-VEHICULAR ACTIVITY (EVA)

Any activity which takes place outside a **spacecraft**, whether in space or on the surface of a **planetary body**; the opposite of **intra-vehicular activity** (IVA). An EVA in space is colloquially termed a 'spacewalk' [figure E7].

[See also **spacesuit**.]

EXTRA-VEHICULAR MOBILITY UNIT (EMU)

An alternative term for a **spacesuit**; more generally extended to include the **manned manoeuvring unit** (MMU) and emergency **life support** and rescue equipment (e.g. the **personal rescue sphere**).

EXTRA-VEHICULAR PRESSURE GARMENT

See **spacesuit**.

F

FACE-SKIN

A thin sheet of material bonded to the surface of a spacecraft structural panel, **antenna** reflector, etc; an integral part of a **honeycomb panel** or the conductive surface of a reflector, for example. Typically made from an alloy of **aluminium**, **titanium**, etc.
[See also **materials**.]

FACSIMILE

A telegraphic system whereby textual or pictorial documents are scanned photoelectrically, producing a signal which can be transmitted via **transmission line**s, **microwave link**s, **satellite communications** links, etc, and reproduced at a distant receiver. Often abbreviated to 'fax'.

FAIRING

Any component part of a **launch vehicle** designed to reduce **drag** by smoothing the flow of air around the vehicle. A section of a vehicle covered by a fairing is said to be 'faired in'.
[See also **skin**, **skirt**, **shroud**, **ogive**, **airframe**, **inter-stage**, **aerodynamics**.]

FATIGUE

(i) The weakening or deterioration of a material under load (e.g. 'metal fatigue'), especially when subjected to cyclic stresses such as vibration.

(ii) Physical or mental exhaustion due to exertion (e.g. 'pilot fatigue').

[See also **fracture mechanics**, **load path**, **vibration table**, **stress—corrosion cracking**.]

FAX

A colloquial abbreviation for **facsimile**.

FCC

An acronym for Federal Communications Commission, the regulatory body for **telecommunications** within the United States and between the USA and foreign **carrier**s.

FDM

See **frequency division multiplexing**.

FDMA

See **frequency division multiple access**.

F/D RATIO

The ratio between the focal length, *F*, and the diameter, *D*, of an **antenna**, or a lens or mirror of an optical system.

FEED

A device designed to illuminate the surface of an antenna reflector when transmitting an RF signal, and to collect radiation reflected from the antenna when receiving. The most common type is the **waveguide** termination known as a **feedhorn**, but various types of 'wire antenna' are also used [see **antenna**].

FEEDER LINK

An alternative term for **uplink**.

FEED(ER) LOSSES

RF losses in a **feed** system—see *I²R* **loss**.

FEEDHORN

A flared and open-ended termination to a **waveguide** designed to illuminate the surface of an **antenna** reflector when transmitting an RF signal, and to collect radiation reflected from the antenna when receiving. Horns can be rectangular in cross section ('E-plane' or 'H-plane' horns), square-ended to accommodate both E- and H-plane waves ('pyramidal'), or circular. Feedhorn is often abbreviated to 'feed', although this term is sometimes used to describe the feedhorn *and* the length of waveguide to which it is attached; it also includes other types of **feed**.

FIGURE OF MERIT (G/T)

A parameter which defines the quality of a receiving installation (**antenna** and **receiver**), otherwise known as G/T, where G is the **gain** (dB) of the antenna and T is the receive system **noise temperature** (dBK). Both quantities are referred to the **feed** flange or LNA input. The units of G/T are dB K^{-1}.

FILAMENT WINDING

A method of manufacturing low-density spacecraft structures, which involves winding a continuous length of resin-impregnated carbon—fibre, or similar material, around a former to produce a hollow body, such as a propellant tank [figure A5] or payload support structure. Filament winding is a finer version of '**tape wrapping**'.

FILTER

An electrical or **microwave** device designed to allow a selected range of signal frequencies to pass, while obstructing those outside the range. The wanted signals are said to be in the 'passband'; the unwanted signals are in the 'stopband'. In addition to confining the **bandwidth** of the signals entering a communications system, a filter reduces the possibility of interference between transmitted signals (but see **mixing products** and **spurious signal**).
[See also **low-pass filter, high-pass filter, band-pass filter, channel filter, input filter, output filter**.]

FIN

(i) An aerodynamic appendage fixed to the body of a **launch vehicle**, typically at the base of the first stage, which provides stability during flight.
[See also **airframe**.]

(ii) A projecting rib or plate which increases the effective surface area of a **radiator** and enhances heat dissipation.
[See also **heat rejection, heat sink**.]

FINITE-ELEMENT MODELLING

A computer-based technique for predicting the behaviour of a structure under mechanical loads. Otherwise known as finite-element analysis.

The structure is divided into a finite number of separate parts, or 'elements', small enough to be assigned realistic values for mechanical loading, etc, but not too numerous that the integration process becomes unwieldy. The technique allows the behaviour of a structure under stress to be accurately assessed at the design concept stage, which can lead to improvements in the final manufactured structure.

FIRING ROOM

See **launch control centre (LCC)**.

FIRST SPACE VELOCITY

See **escape velocity**.

FIXED CONDUCTANCE HEAT PIPE (FCHP)

See **heat pipe**.

FIXED-SATELLITE SERVICE (FSS)

A **satellite communications** system which uses **earth stations** at specified fixed points. This service is restricted to a relatively small set of user organisations and is not directly accessible to the general public. The earth stations at one end of the link tend to be those with very large **antennas**, trunk stations or **gateway stations**, while those at the **user** level can be of any size down to about 2 m in diameter (depending on the specific parameters of the system). In the formal **ITU** definition, FSS may include satellite-to-satellite links or **inter-satellite links**, which are included in the inter-satellite service (ISS).
[See also **MSS** and **BSS** for mobiles and broadcasting respectively.]

FIXED SERVICE STRUCTURE

See **service structure**.

FLAME BUCKET

See **flame deflector**.

FLAME DEFLECTOR

An obstruction designed to intercept a **launch vehicle**'s exhaust gases and deflect them away from the structure of the **launch pad**, the ground and the vehicle itself. Designs vary in relation to the size and **thrust** of the vehicle, from relatively small devices fixed to the top surface of the pad to very large deflectors, sometimes called 'flame buckets', beneath the pad. The larger devices are typically deluged with water during a launch to constrain the temperature of the structure and avoid damage.
[See also **flame trench, sound suppression system**.]

FLAME INHIBITOR

A material placed between the **motor case** and the **propellant** in a solid rocket motor which protects parts of the case that might otherwise be damaged (even burned through) before propellant combustion is complete. Also known as a 'propellant liner'—see **rocket motor**.

FLAME TRENCH

A channel or gully beneath a **launch pad** which carries a **launch vehicle**'s exhaust products away from the pad. It usually contains a **flame deflector**.

FLAT ABSORBER

See **surface coatings**.

FLAT REFLECTOR

See **surface coatings**.

FLIGHT

(i) Of a **spacecraft**, any journey in space: a **spaceflight**.

(ii) Of a **launch vehicle**, the time interval between **lift-off** and the completion of its mission—when its **payload** has been injected into an **orbit** or **trajectory**.

(iii) Of an **aerospace vehicle**, any journey in space or the portion of its journey which takes place in the atmosphere.

[See also **launch**.]

FLIGHT CONTROLLER

A person in a **launch control centre** with responsibility for the **flight** of a **launch vehicle**.

FLIGHT DECK

The part of an aircraft or **manned spacecraft** from which the **flight** is controlled by a pilot and/or co-pilot. In the case of the American Space Shuttle **orbiter**, the 'aft flight deck' also allows a view into the **payload bay** and houses the controls for the **remote manipulator system** (RMS). [See also **mid-deck**.]

FLIGHT HARDWARE

Any equipment which leaves the surface of the Earth, or another **planetary body**, in the course of a **flight**; a **launch vehicle**, a **spacecraft** or any of their constituent **subsystem**s or equipment.
[See also **ground support equipment (GSE), hardware**.]

FLIGHT MODEL

The version of a spacecraft built specifically for flight, as opposed to some phase of ground testing (e.g. **structure model, thermal model**). The flight model undergoes tests slightly less severe than the other models (e.g. narrower temperature range).
[See also **thermal-vacuum chamber**.]

FLIGHT PATH

The course of a **spacecraft** or **launch vehicle** through a planetary atmosphere; also called a **trajectory**. The term 'trajectory' is also used for a body moving outside an atmosphere, whereas 'flight path' tends to be confined to atmospheric flight.
[See also **flight, glide path**.]

FLIGHT READINESS FIRING (FRF)

The final test firing of a **launch vehicle**'s main **propulsion system** prior to an actual **launch**, particularly of the American **Space Shuttle main engine**s.

Such tests, which are conducted with the vehicle held down on the **launch pad**, concern only **liquid propellant** engines since *they* can be tested for a short period compared with the flight **burn** duration, shut down, and then refuelled on the pad; **solid propellant** motors, once ignited, must burn for the full duration, until their propellant is exhausted, and cannot be refuelled on the pad.
[See also **static firing, test stand, hold-down arm**.]

FLOTATION BAG

An inflatable device in the nose cone of the American **Apollo** spacecraft, designed to inflate on **splashdown**, right the spacecraft if it landed 'upside down' and keep it afloat until the **recovery** services arrived.

FLOTATION COLLAR

An inflatable collar attached by divers to **Gemini** and **Apollo** spacecraft after **splashdown** to keep the spacecraft upright and to stop it sinking.

FLOW SEPARATION

In general, the detachment of the boundary of any moving fluid from the surface by which its flow is guided (e.g. the separation of the air flow from an aerodynamic surface).

In **rocket** propulsion, the separation of the gas flow from the inside of an exhaust **nozzle**. If a nozzle is designed for optimum efficiency at high altitude (i.e. in vacuum or near-vacuum conditions), air pressure at lower altitudes will compress the exhaust **plume** causing flow separation, thus reducing the **expansion ratio** and the efficiency of the device. Conversely, a nozzle designed for optimum efficiency at low altitude performs less efficiently at high altitude. The **exit cone** of a **launch vehicle** nozzle is therefore specifically designed to operate with optimum efficiency over the range of altitude flown by its parent **stage** . [See also **plug nozzle**.]

FM

An acronym for (i) **frequency modulation** and (ii) **flight model**.

FOOT RESTRAINT

See **restraint**.

FOOTPRINT

(i) The area within which a spacecraft is expected to land. Usually depicted on a map as an ellipse with its major axis aligned with the direction of travel.

(ii) The projection of a satellite's antenna **beam** on the Earth's surface [figure F1]. Usually depicted as a contour map of **radiated power** received at the surface, with the **peak** power at the centre of the footprint. The footpring or 'beam shape' may be circular, elliptical or a complex shaped beam, depending on the design of the originating antenna.

'Footprint' is a descriptive and inexact term and bears no mathematical relationship to the more closely defined **service area** and **coverage area**.
[See also **beamwidth, multiple beam, equivalent isotropic radiated power (EIRP), power flux density (PFD)**.]

FORWARD ERROR CORRECTION (FEC)

A process for detecting and correcting errors in a communications link withouth retransmission. The information is divided into blocks and labelled with 'checking information' which is recalculated after transmission and compared with the original. Any errors are corrected on this 'forward' path rather than requesting retransmission on the return path.

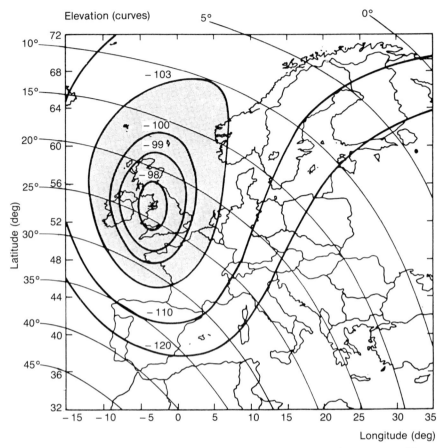

Figure F1 The **footprint** devised at the 1977 **WARC** conference on **direct broadcasting by satellite** for the United Kingdom. The contours indicate levels of **power flux density (peak** is 97.7 dBW m^{-2}. [British Aerospace]

FRACTURE MECHANICS

The science of predicting the mechanisms of material failure and designing against them.
[See also **fatigue, stress (engineering)**.]

FRAME OF REFERENCE

A known arrangement of planes or curves used to determine the position of a point in space. The most common frame of reference is that of cartesian geometry, characterised by three mutually perpendicular (orthogonal) planes.
[See also **spacecraft axes, degree of freedom**.]

FREEDOM

The name given to the US/International **Space Station**.

FREE-DRIFT STRATEGY

A satellite **station-keeping** method which increases the north—south **dead-band**, allowing the satellite to drift north and south of the equatorial plane. It reduces the **propellant** requirement for north—south station keeping.

This strategy can be applied if the **earth station**s can accommodate a large excursion in latitude [see **box**] by **tracking** the satellite. Under the right conditions, a satellite can be given an **orbital inclination** at **beginning of life** which causes it to describe a 'figure 8' completely filling its box. The perturbing effects of **luni—solar gravity** decrease the inclination, reducing the size of the figure 8 to a minimum about halfway through the satellite **lifetime**, and then increase it until a maximum is reached at **end of life**. The operational variations between satellites, which may call for smaller boxes or longer lifetimes, often mean that a combination of free drift and N—S station keeping is required.

FREE FALL

See **weightlessness**.

FREE-FLYING PALLET

See **pallet**.

FREE-RETURN TRAJECTORY

A **trajectory** which takes a spacecraft out to, around and back from a **planetary body** using only the gravitational field of that body, i.e. without the spacecraft expending energy. This trajectory was used, for example, during the **Apollo** lunar programme, for both the Apollo 8 and 13 missions.
[See also **trans-lunar trajectory**, **ballistic trajectory**.]

FREE SPACE

A term used to describe a volume of 'space' which is free of anything which might perturb or affect the system under discussion: matter, electromagnetic radiation, gravitational fields, etc. Free space is therefore something of an approximation. For radio communications between **earth station**s and **satellite**s in orbit, the distance of free space is used in the calculation of signal strength [see **free-space loss**].

FREE-SPACE LOSS

A loss in power density of a radiated telecommunications signal due simply to the distance of **free space** between **transmitter** and **receiver**. This major source of loss is inherent in a spacecraft communications system due to the remote nature of the spacecraft.

For example, for a simple **antenna** with some degree of **directivity**, radiated power is spread into a cone in accordance with the antenna **beamwidth** (like a huge searchlight beam). In the case of an earth station with an antenna beamwidth of 0.25 degrees, the **beam** is over 150 km in diameter at the height of the **geostationary orbit** (36 000 km). The power level at the satellite is reduced by the same ratio as that between the area of the beam at that height and the surface area of the ground antenna, which is around 10^{10}:1. In these terms space loss is otherwise known as 'spreading loss', but since beamwidth is frequency dependent, it is common (and convenient) to include a wavelength term in the calculation of free-space loss.

Thus the loss in signal power in a communications link between an **earth station** and a **satellite**, due to the signal's passage through free space, may be calculated using the following expression:

$$20\log_{10}\left(\lambda/4\pi R\right)$$

where λ is the **wavelength** in m and R is the distance in m between the satellite and the earth station (the depth of the atmosphere is usually ignored in this calculation but is taken into account in a calculation of **atmospheric attenuation**).

FREQUENCY

The number of oscillations per second of an electromagnetic wave. Frequency is measured in hertz (Hz) or multiples of hertz. Satellite communications frequencies are predominantly in the GHz (gigahertz: 10^9 Hz) and MHz (megahertz: 10^6 Hz) bands. See also **frequency bands**.

The relationship between frequency and **wavelength** is $f = c/\lambda$, where f is frequency (Hz), λ is wavelength (m) and c is the speed of light (3×10^8 m s^{-1}).

FREQUENCY ALLOCATION

The process by which the International Telecommunications Union (**ITU**) allocates either exclusive or shared **bandwidth** to telecommunications services and all other users of the **radio-frequency** spectrum. For frequency allocation, the ITU has divided the world into three regions: Region 1 includes Europe, Africa, the USSR and Mongolia; Region 2 the Americas and Greenland; and Region 3 Asia, Australasia and the Pacific. Frequency allocation is also known as frequency assignment or allotment. [See also **frequency bands**.]

FREQUENCY ALLOTMENT

See **frequency allocation**.

FREQUENCY ASSIGNMENT

See **frequency allocation**.

FREQUENCY BAND

A continuous range of frequencies between two limits. The **frequency bands** used by all communications services, and those protected for the benefit of astronomical observations, are allocated by the International Telecommunications Union (**ITU**) at its World Administrative Radio Conference (**WARC**) meetings.
[See also **frequency allocation**.]

FREQUENCY BANDS

The **radio spectrum** has been divided into bands by the **ITU**, as shown in table F1.

Table F1 Division of radio spectrum into bands by the ITU.

Symbol†	Frequency range†† (lower limit exclusive, upper limit inclusive)	Corresponding metric subdivision (note: not widely used)
VLF	3 − 30 kHz	Myriametric waves
LF	30 − 300 kHz	Kilometric waves
MF	300 − 3000 kHz	Hectometric waves
HF	3 − 30 MHz	Decametric waves
VHF	30 − 300 MHz	Metric waves
UHF	300 − 3000 MHz	Decimetric waves
SHF	3 − 30 GHz	Centimetric waves
EHF	30 − 300 GHz	Millimetric waves
	300 − 3000 GHz	Decimillimetric waves

† Key: V = very; U = ultra; S = super; E = extra; L = low; M = medium; H = high; F = frequency.

†† Prefixes are those assigned by the ITU, i.e. k = kilo (10^3), > 3 kHz, ≤ 3000 kHz; M = mega (10^6), > 3 MHz, ≤ 3000 MHz; G = giga (10^9), > 3 GHz, ≤ 3000 GHz; T = tera (10^{12}), > 3000 GHz, i.e. centimillimetric, micrometric and decimicrometric waves.

Satellite communications frequencies are mainly in the SHF band, but UHF and VHF are also used. For many communications applications the sub-bands are given letter designations. The following example is the IEEE Radar Standard 521:

L-band :   1–2 GHz
S-band :   2–4 GHz
C-band :   4–8 GHz
X-band :   8–12 GHz
Ku-band: 12–18 GHz
K-band :   18–27 GHz
Ka-band: 27–40 GHz

Standard 521 also defines 40 − 100 GHz as the millimetre waveband (mm-wave), and 0.3 − 1 GHz as UHF (at variance with the ITU definition).

These designations are intended as a guide to common usage in the satellite communications industry and should not be taken as definitive for all applications. Over the years, several frequency-band standards have evolved for use in terrestrial radio, radar and spacecraft communications, often differing according to which side of the Atlantic they were devised. Thus readers may also come across, amongst others, P-band (0.23–1 GHz), R-band (1.7–2.6 GHz), H-band (3.95–5.85 GHz), and designations which differ in that Ku-band is shown as J-band and Ka-band as Q-band. It is also common for the 1–300 GHz band to be defined as the **microwave** band.

Although **frequency allocation** is an extremely complex subject, it can be said that each particular band is dominated by certain services: communications satellites operate at C-band, Ku-band and, increasingly, K- and Ka-band; X-band is largely reserved for military communications; L-band is used for the **mobile-satellite service**; and S-band is used, amongst other things, for satellite **telemetry** and **telecommand**. Frequency combinations such as 4/6 GHz, 12/14 GHz, 12/18 GHz and 20/30 GHz are commonly used: these refer to approximate **uplink** and **downlink** frequency bands for various satellite services. The uplink is always the higher figure, since the greater losses (due to **atmospheric attenuation**) at higher frequencies can be more readily overcome by higher-power earth station transmitters (cf increasing the spacecraft HPA power).

FREQUENCY COORDINATION

The process which ensures the simultaneous operation of both **satellite** and terrestrial communications systems without **interference**. The case for the frequency coordination of a satellite or earth station is usually made to the International Frequency Registration Board (**IFRB**) via the national radio regulatory administration, the local coordinating body.

The **frequency** coordination of a satellite earth station, for example, begins when a new site is proposed. The procedure entails the submission of details of **channel** frequencies, **orbital position** of the target satellite, geographical location of the intended station, a polar diagram of the **antenna radiation pattern** and other parameters such as **transmitter** power and data format. An **interference analysis** is performed to decide whether the earth station will interfere with existing communications links, and vice versa. A similar range of criteria is applied to the frequency coordination of a satellite.
[See also **frequency allocation.**]

FREQUENCY DIVISION MULTIPLE ACCESS (FDMA)

A coding method for information transmission between a potentially large number of users, whereby each user is allocated a relatively narrow section of the total transponder **bandwidth** (i.e. at slightly different frequencies) to which they have continual access. The satellite **transponder** can thus be accessed by a relatively large number of users, thereby making efficient use of the total available bandwidth. FDMA was the first multiple access scheme to be developed and was first used on **Intelsat** satellites.
[See also **time division multiple access.**]

FREQUENCY DIVISION MULTIPLEXING (FDM)

In **telecommunications**, the process of placing more than one **signal** on a **carrier**, whereby each signal is given its own **subcarrier** frequency. Each subcarrier is modulated independently, then each modulates the main carrier. Thus the main carrier conveys all the information signals (or **channels**) at the **frequency** of operation of the **transmitter**. Frequency multiplexing allows continuous **data** transmission from all channels, unlike **time division multiplexing**.
[See also **frequency modulation.**]

FREQUENCY HOPPING

The practice of switching a radio transmission between different radio frequencies in order to deter unauthorised reception, used mainly in military communications systems.

FREQUENCY MODULATION (FM)

A transmission method using a modulated **carrier** wave, whereby the **frequency** of the carrier is varied, around its nominal or 'centre frequency', in accordance with the **amplitude** of the lower frequency input signal. The amplitude of the carrier remains unchanged.

FM, which came into practical use just before the Second World War, was developed to improve on **amplitude modulation** (AM) which suffers from **noise**. Practically all natural and man-made radio noise consists of amplitude disturbances which, in the radio receiver, are indistinguishable from the modulation of AM. Raising the transmission power improves the **signal-to-noise ratio**, but is costly and becomes ineffective over long distances. Frequency modulation, by contrast, is unperturbed by amplitude noise since the amplitude remains nominally constant; any amplitude modulation can be removed by means of a 'limiter' in the receiver. Terrestrial AM broadcasting was also characterised by a lack of fidelity, but this was mainly due to the historical allocation of channel **bandwidth**s too narrow to encompass the complete range of frequencies detectable by the human ear. FM removed this limitation by reproducing the full audio band (from 20 Hz to 15 kHz) at VHF where **frequency space** for a wider bandwidth is available.

[See also **modulation, phase modulation (PM), pulse code modulation (PCM), delta modulation (DM), QPSK (quadrature phase shift keying).**]

FREQUENCY REUSE

The practice of using the same **radio frequency** more than once in the same satellite **beam** or **coverage area**. The aim of this practice is to make the most efficient use of the limited resource of **frequency space**. In one form it involves the use of opposite **polarisation**s (*polarisation frequency reuse*) [see **dual-gridded reflector**]. In another form (*spatial frequency reuse*) the same frequencies are used in more than one **spot beam**. A more advanced form utilises **antenna**s which transmit three or four different frequencies into adjacent spot beams arranged in a pattern repeated many times across the coverage area. A practical system of this type requires a complex switch network on the satellite to direct the communications **traffic** into the correct beam via the correct **feedhorn**.

FREQUENCY SEPARATION

The allocation of different radio frequencies to communications services which, due to their physical proximity, would otherwise interfere with each other. Essentially the same as **channel separation**. See **interference**.

[See also **radio frequency (RF).**]

FREQUENCY SOURCE

See **local oscillator**.

FREQUENCY SPACE

In theoretical terms, the concept that there is an available commodity called '**frequency**', a usable but not unlimited asset. The collective term for this commodity is frequency space. On a more practical level the term refers to a range of frequencies (if continuous, a **band** of frequencies) in the sense of 'space in the **frequency band**'.

FREQUENCY SPECTRUM

An alternative name for the **electromagnetic spectrum**; the spectrum in terms of **frequency** as opposed to **wavelength**. The frequency spectrum contains, by definition, the whole range of frequencies; the radio-frequency part of the frequency spectrum is more confined (10 kHz–300 GHz)—[see **radio frequency (RF)**]. It is also called the radio-frequency spectrum or radio spectrum.
[See also **frequency bands**.]

FRISBEE EJECTION SYSTEM

A method of releasing spacecraft from the **payload bay** of the American **Space Shuttle** specifically developed for the **spin-stabilised** Leasat and Intelsat VI satellite programmes (named after the 'Frisbee' recreational product, a light plastic disc thrown with a spinning motion). The spacecraft is mounted in a U-shaped **cradle**, with its spin axis parallel to the Shuttle's longitudinal axis. An attachment point on one side of the bay acts as a pivot point and an ejection spring on the other side, when released, causes the spacecraft to rotate about the pivot. This creates a similtaneous translation and rotation about the spacecraft's centre of mass (typical values: 0.36 m s^{-1} and 2 rpm), and the spacecraft appears to roll up an imaginary ramp out of the payload bay [figure F2].

FRONT END

See **receiver**.

FSS

See **fixed-satellite service**.

FUEL

The component of a rocket **propellant** which is burned or 'oxidised' in a chemical reaction with an **oxidiser** to produce **thrust** (e.g. **liquid hydrogen, kerosene, polybutadiene, hydrazine,** MMH, UDMH).
[See also **solid propellant, liquid propellant**.]

FUEL BUDGET

See **propellant budget**.

FUEL CELL

A spacecraft power source which generates energy by the electrochemical combination of two fluids, usually hydrogen and oxygen. The conversion of chemical energy to electrical energy is the opposite of the electrolysis of water, itself a by-product of the fuel cell reaction. Otherwise known as the 'Grove cell' after the British physicist Sir William Robert Grove (1811–1896) who invented the fuel cell in 1839.

The fuel cell has been used in American manned spacecraft since the Gemini programme, but the requirement for consumables limits the duration of the mission and makes the fuel cell highly unsuitable for unmanned spacecraft with **design lifetime**s of several years.

[See also **power, solar cell, radioisotope thermoelectric generator**.]

Figure F2 A Syncom IV/Leasat **communications satellite** being **deploy**ed from the **Space Shuttle** using the **frisbee ejection system**. [NASA]

G

G FORCE

An abbreviation for 'gravitational force', used particularly with regard to the force experienced by an **astronaut** in an accelerating or decelerating spacecraft, and by aircraft test-pilots, etc. See **gravity**.

GaAsFET

An abbreviation for gallium arsenide field effect transistor, a semiconductor device widely used in **satellite communications** equipment.

GAIN

The ratio of **output** power to **input** power of an **amplifier**, usually measured in decibels (dB). See **antenna gain**.

GAITER

A flexible sealing-collar, fixed between a **rocket motor** or **rocket engine** and the surrounding structure of a **propulsion bay**, which protects the bay from the effects of heat and **exhaust** gases while allowing the **nozzle** to **gimbal** for **thrust vector** control.

GALILEO

An American planetary exploration **probe** destined to be placed in orbit around the planet Jupiter and release an entry probe into the Jovian atmosphere.

GANTRY

A tower framework which allows access to a **launch vehicle** on its **launch pad**. See **service structure**.

GAS

An acronym for **getaway special**.

GAS GENERATOR

A device used to drive a turbine (e.g. in the **turbopump** of a **rocket engine** or in the **auxiliary power unit**s (APUs) of the American **Space Shuttle**). In the APU, **hydrazine** is decomposed into a gas by passing it over a bed of a granular catalyst (similar to that of the **hydrazine thruster**). The gas is then expanded to decrease its pressure, accelerated to a high velocity and directed by a nozzle at a set of turbine blades to drive an hydraulic pump. In a rocket engine, the turbine drives a turbopump which delivers **propellant** to the engine(s).

GAS JET

A colloquial term for **reaction control thruster**.

GASEOUS OXYGEN ARM

A **swing-arm** on the fixed **service structure** at the American Space Shuttle **launch pad** which supports a device to collect gas vented from the **liquid oxygen** tank within the Shuttle's **external tank**. See **beanie cap**.
[See also **vent and relief valve**.]

GATEWAY STATION

A major **earth station** which handles incoming and outgoing international **telecommunications** traffic—analogous to a 'gateway airport'. Originally the definition was limited to a few **Intelsat** earth stations, characterised by their very large **antenna**s (30 m diameter or larger). Nowadays it refers to earth stations as small as 11 m in diameter operated by any authorised **common carrier**.
[See also **Intelsat standard earth stations**.]

GEMINI

A series of ten American **manned spacecraft** launched in the mid 1960s to continue research into manned orbital flight after the **Mercury** programme and prepare for the **Apollo** lunar programme. Gemini did this mainly by practising **rendezvous** and **docking** techniques, using the **Agena** target vehicle, and proving that **astronaut**s could survive in space for up to two weeks as future Apollo crews would have to.

The first manned mission was Gemini 3, launched 23 March 1965 (mission duration 4 h 53 min); Gemini 6 rendezvoused with Gemini 7 (15 December 1965) and the latter remained in orbit nearly 2 weeks.

The Gemini spacecraft comprised two sections: a **re-entry** module (the capsule itself) and an adapter module. Unlike Mercury, the two-man Gemini **capsule** had two ejection seats instead of an escape tower. The capsule was depressurised for EVA (**extra-vehicular activity**). The adapter module, which was tapered to fit the capsule to the **Titan** II launcher, carried service and **propulsion** subsystems. Power was supplied by **fuel cells** and **attitude control** was accomplished by **thrusters** using MMH and **nitrogen tetroxide**. The adapter module was jettisoned prior to re-entry to expose a **heat shield** based on a glass-fibre honeycomb and an ablative organic compound.

GEO

A common abbreviation for **geostationary orbit**. It can be thought of as an acronym for 'geostationary earth orbit', but this is tautologous since the prefix 'geo' refers to Earth. It is sometimes used as an acronym for geostationary equatorial orbit.

GEOSTATIONARY ARC

The portion of the **geostationary orbit** 'visible' from a particular place on Earth. The observer's latitude and longitude define the length of the arc, and therefore the number of **orbital position**s visible, and its **declination**, which is important in that it governs the degree of **atmospheric attenuation**.

GEOSTATIONARY ORBIT

A **circular orbit**, of radius 42 164 km (26 200 miles) and height (above) the Earth's surface) 35 786 km (22 237 miles), in the same plane as the Earth's equator. Commonly abbreviated to GEO; sometimes GSO (for geostationary satellite orbit or simply geostationary orbit) [see figure G1].

A satellite in this **orbit** has an orbital period equal to the Earth's **sidereal period** of rotation (23 h 56 min 4 s); the Earth's **synodic period** is 24 h. The satellite therefore appears stationary with respect to the Earth—hence 'geostationary'—which means it can be given an **orbital position** related to the line of longitude it is stationed above.

GEO is the orbit used for most **communications satellite**s, partly because small **earth terminal**s, with their relatively wide **beamwidths**, do not need to 'track' the satellite, which decreases the cost of the **ground segment**. In addition, a system of three equally spaced satellites can provide full coverage of the Earth, except for the polar regions where the **elevation angle** is very low. The possible application of this useful orbit was first brought to public attention in the October 1945 issue of 'Wireless World' by the writer Arthur C Clarke. Geostationary orbit is thus sometimes known as the 'Clarke Orbit'.

The term 'geosynchronous' is often heard in place of 'geostationary', but the latter is a subset of the former. See **geosynchronous orbit**. [See also **perturbations**.]

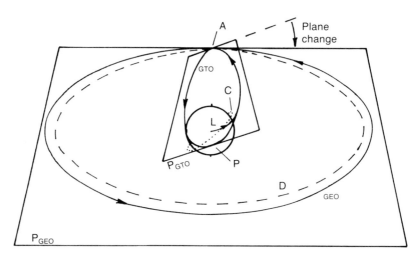

Figure G1 The route to **geostationary orbit** (GEO) via **geostationary transfer orbit** (GTO) and a launch vehicle **trajectory** (L) to GTO. PGEO = plane of GEO, PGTO = plane of GTO [see **plane change**]; A = apogee; P = perigee; D = **drift orbit**; C = circular **low Earth orbit** (e.g. as attained by **Space Shuttle**).

GEOSTATIONARY TRANSFER ORBIT (GTO)

An elliptical Earth orbit used to transfer a spacecraft from a low altitude **orbit** or flight **trajectory** to **geostationary orbit** (GEO) [see figure G1]. The orbit has its **apogee** at geostationary height (36 000 km): when the spacecraft reaches this point, its **apogee kick motor** is fired to inject it into GEO.

    Expendable launch vehicles (ELVs) usually inject their **payloads** directly into GTO using their third (or fourth) **stage**. The American **Space Shuttle**, in contrast, can only deliver a payload to a **low Earth orbit** or **parking orbit**, from where an additional rocket motor (a **perigee kick motor** or **upper stage**) boosts it to geostationary orbit (via a GTO). The geostationary transfer orbit is a derivative of the minimum energy **Hohmann transfer orbit**.

[See also **parasitic station acquisition**.]

GEOSYNCHRONOUS ORBIT

An **orbit** whose period of rotation is some multiple or submultiple of the Earth's rotational period; an orbit in which the orbiting body is synchronised with the Earth.

    A **satellite** in such an orbit will pass over the same point on the Earth at a given time (or times) each day. For example, a satellite in an **equatorial orbit** with a period of 12 hours is synchronous in that it passes over the same point twice each day. See **geostationary orbit**.

[See also **heliosynchronous orbit**.]

GETAWAY SPECIAL (GAS)

A colloquial term for the small self-contained payload (SSCP); a **payload** designed to be enclosed in a cylindrical canister mounted on the side of the American **Space Shuttle**'s **payload bay**. The getaway special offers relatively inexpensive access to the space environment for non-commercial payloads, typically from school and university groups.

GHz

Abbreviation for gigahertz—see **hertz, frequency bands**.

GIMBAL

(i) A device used to support a **gyroscope** which gives its spin axis a **degree of freedom**. See **inertial platform**.
[See also **strapdown gyro system**.]

(ii) A device on which a **rocket engine** can be mounted to provide angular movement in one or two dimensions, used to steer the vehicle. (NB: also used as a verb: 'to gimbal'.)

GIOTTO

A European Space Agency (**ESA**) spacecraft designed to investigate Halley's Comet; **encounter** occurred 13 March 1986.

GLIDE PATH

The course of a **spacecraft**, particularly an unpowered winged vehicle, which generates **lift** in a planet's atmosphere to enable a controlled landing to be made. Since the path invariably subtends an angle to the planet's surface, it is also known as a 'glide slope'.
[See also **flight path, trajectory, lifting surface, aerospace vehicle**.]

GLIDE SLOPE

See **glide path**.

GLITCH

Jargon: any spurious pulse experienced in the operation of an electrical circuit. The term has been extended to include any unexpected occurrence in any pre-planned sequence of events (e.g. 'a glitch in the **countdown** sequence').
[See also **spurious signal**.]

GLOBAL BEAM

A **beam** formed by a **satellite** communications **antenna** which provides coverage to all parts of the globe 'visible' from the satellite. A satellite in

geostationary orbit can communicate effectively with about one third of the Earth's surface, hence global communications satellite networks feature at least three satellites, usually stationed over the Atlantic, Pacific and Indian Oceans.
[See also **beamwidth, hemispherical beam, zone beam, spot beam, multiple beam.**]

GODDARD SPACE FLIGHT CENTER (GSFC)

See **NASA**.

GORIZONT

The name given to a series of Soviet geostationary **communications satellite**s.

GRACEFUL DEGRADATION

Jargon: a gradual decrease in the performance or **capacity** of a **system**, as opposed to a sudden failure.
[See also **degradation, soft fail.**]

GRAIN

See **propellant grain**.

GRAPHITE EPOXY

See **carbon composite**.

GRAPPLE FIXTURE

A device attached to a spacecraft which allows it to be handled by the American Space Shuttle's **remote manipulator system** (RMS) [see figure R3]. Typically a shaft with a broadened tip, which can be 'grappled' by three 'snare wires' within the manipulator arm's **end effector**. It also includes an arrangement of three 'guide ramps' which mate with compatible 'key-ways' in the end effector to secure the spacecraft to the RMS.

GRAVEYARD ORBIT

A colloquial term for an **orbit** beyond **geostationary orbit** to which satellites are boosted at the end of their lives, in order to leave room for their replacements. It is common to allocate an amount of thruster **propellant**, which would otherwise be used for **station keeping**, for this activity.
[See also **end of life, propellant budget.**]

GRAVITATIONAL ANOMALY

A local irregularity in the gravitational field of a **planetary body** — see **mascon**.
[See also **perturbations**.]

GRAVITATIONAL ATTRACTION

See **gravity**.

GRAVITATIONAL BOOST

A technique in which the gravitational field of a **planetary body** is used to accelerate and re-direct a **spacecraft** on an interplanetary **trajectory**; also known as a 'gravitational sling-shot' technique [figure G2]. For example, the planet Jupiter was used to boost the **Pioneer** and **Voyager** spacecraft on their trajectories to the outer Solar System.

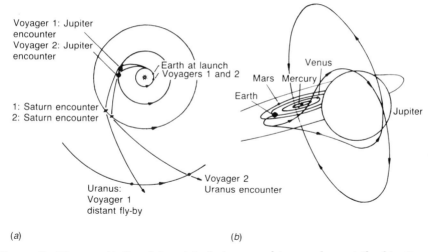

(a)   (b)

Figure G2 The **gravitational boost** technique used to supplement the kinetic energy of a **spacecraft** on an interplanetary **trajectory**; (a) **Voyager** 1 and 2 trajectories to Saturn and beyond; (b) double sling-shot manoeuvre originally planned for two international solar polar mission (ISPM) spacecraft, later reduced to one spacecraft renamed Ulysses.

GRAVITATIONAL CONSTANT

See **gravity**.

GRAVITATIONAL FIELD

See **gravity**.

GRAVITATIONAL PERTURBATION

A disturbance (of a spacecraft **orbit** or **trajectory**) caused by the gravitational attraction of the planetary bodies. See **perturbations**.

GRAVITY

The force of attraction which exists between any two bodies. Also called 'gravitation'.

The gravitational force, F, between two bodies is proportional to the product of their masses (M and m) and inversely proportional to the square of the distance, r, between them. It is given by the expression

$$F = \frac{G\,M\,m}{r^2}$$

where G is the gravitational constant, 6.67×10^{-11} N m^2 kg^{-2}.

Gravity is an inherent property of matter. Any conglomeration of matter, however small, creates a gravitational attraction which accelerates any other body towards it—this is the acceleration due to gravity, which on Earth is approximately 9.81 m s^{-1} ('1g'). The conceptual 'field of infuence' of the gravitational force is termed the 'gravitational field'.

[See also **microgravity**, **perturbations**, **gravitational anomaly**, **gravity gradient stabilisation**.]

GRAVITY GRADIENT STABILISATION

A method of stabilising the attitude of a spacecraft, used predominantly in **low Earth orbit**, which makes use of the decrease in the Earth's gravitational field strength with distance from its centre. The difference in gravitational attraction between the two ends acts to maintain the body's orientation.

Its use requires that the spacecraft body should be as long as possible with its mass concentrated about an axis aligned with the Earth's centre (the **yaw axis**). Although the body can rotate about the yaw axis, its orientation is restricted and this method of **attitude control** is useful predominantly for the simplest satellites [see **tethered satellite**]. The method could, however, be used for future **space platform**s.

GREGORIAN REFLECTOR

A style of dual reflector **antenna** using a concave ellipsoidal **subreflector** and paraboloidal main reflector. The **feedhorn**, subreflector and main reflector are coaxial [see figure A2]. This geometry was first described for optical astronomical telescopes in 1663 by James Gregory, a Scottish mathematician and astronomer, but due to the difficulty of grinding glass into accurate curves, the first practical example was constructed several years later by Robert Hooke who presented it to the Royal Society.

[See also **Cassegrain reflector, Newtonian telescope, Schmidt telescope, Coudé focus, space telescope**.]

GROUND SEGMENT

The terrestrial part of a **satellite**-based communications system or any communications link with a spacecraft. The term is currently synonymous with **earth segment**, but permanent spacecraft communications installations on any planetary surface could be termed 'ground segment'.

[See also **space segment**.]

GROUND SPARE

A **satellite** stored on Earth which is available to replace a defective satellite in **orbit** or enhance an existing satellite system. The value of such a spare depends on the capability to launch the satellite in time to replace a defective spacecraft.

[See also **in-orbit spare, operational, pre-operational**.]

GROUND STATION

An installation on the Earth comprising all the equipment necessary for communications with a **spacecraft**. See **earth station**.

GROUND SUPPORT EQUIPMENT (GSE)

Any ground-based equipment designed to support a spacecraft mission. It includes non-flight **hardware** for ground handling, servicing, inspecting and testing flight hardware, and **software** for launch checkout and in-flight support [figure G3]. The definition does not include land or buildings but may include physical equipment supports.

GSE is often divided into MGSE and EGSE (pronounced 'megsy' and 'egsy') for mechanical, and electrical ground support equipment. Other similar terms are also used (e.g. PGSE for payload GSE).

GROUND TRACK

The projection of the **flight path** of a vehicle on the surface of a planet [figure G4]. If the vehicle is in orbit, it is also known as an **orbital track**.

Figure G3 The Olympus **communications satellite** mounted on its handling trolley, part of the **ground support equipment**. [ESA] [See also figure L1]

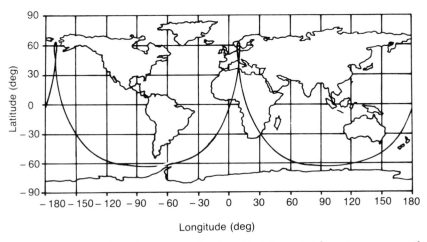

Figure G4 The **ground track** for a satellite in a **Molniya orbit** [see also figure M4].

GSE

See **ground support equipment**.

GSO

See **geostationary orbit**.

G/T

See **figure of merit**.

GTO

See **geostationary transfer orbit**.

GUARD-BAND

A band of frequencies separating two **channels** to guard against **interference**.
[See also **channel isolation**.]

GUIANA SPACE CENTRE

The **launch site** of the European **Ariane** launch vehicle, located at approximately 5°N, 53°W near Kourou, French Guiana; also known by the French acronym CSG (Centre Spatiale Guyanais) [figure G5]. The centre was established by the French space agency **CNES** in 1964 to launch **sounding rocket**s and was later used by **ELDO** (European Launcher Development Organisation) for the **Europa** launch vehicle. In 1976 the European Space Agency (**ESA**) chose the site for Ariane.
[See also **ELA**.]

GUIDANCE (INERTIAL)

See **inertial platform**.

GYRO

An abbreviation for **gyroscope**.

GYROSCOPE

A device containing a spinning disc or 'fly-wheel' mounted on a system of gimbals such that it maintains a fixed orientation irrespective of the motion of the supporting structure or vehicle. The gyroscope remains fixed in **inertial space**; its **frame of reference** is the 'celestial sphere'.

Gyroscopes are used for the detection of rotation relative to inertial space in both **launch vehicles** and **spacecraft**. See **inertial platform**.

Figure G5 An **Ariane** launch vehicle on its **mobile launch platform** at **Guiana Space Centre**. [ESA]

GYROSCOPIC STIFFNESS

The type of stability provided by a **gyroscope**, which tends to maintain the orientation of its spin axis. This property is used, amongst other things, to stabilise **three-axis-stabilised** spacecraft—see **momentum wheel**.
[See also **inertial platform**.]

H

H (LAUNCH VEHICLE)

A three-stage Japanese **launch vehicle** developed in the 1980s. The H-I variant, which first flew in August 1986, uses **liquid oxygen** (LOX) and **kerosene** in its first stage, LOX/**liquid hydrogen** (LH$_2$) in its second and a **solid propellant (HTPB)** in its third. Its first stage **thrust** is augmented by nine solid (**CTPB**) **strap-on** boosters (the same stage and strap-ons are used for the N-II booster [see **N**]). The H-I's **payload** capability to **geostationary orbit** (GEO) is 550 kg. A second variant, the H-II (using LOX and LH$_2$ for its two stages and two large HTPB/**AP**/aluminium **solid rocket boosters**) has been designed to deliver 2000 kg to GEO.

HABITABLE MODULE

A part of a modular **manned spacecraft** (e.g. a **space station**) designed for human habitation, as opposed to **propellant** storage, **power** generation, etc; a spacecraft **module** which incorporates a **life support system**.

HALF-POWER BEAMWIDTH

The width of the beam of radiation formed by a communications **antenna**, measured in degrees, between the points where the radiated **power** is half of its **peak** value. See **beamwidth**.

HALF-TRANSPONDER

Half of the frequency **bandwidth** available to the user in a **transponder**. Service providers (e.g. **Intelsat**, **Eutelsat**, etc) often deal in half-transponders, since not all users require the bandwidth available in a whole transponder.

In **satellite communications**, 'half-transponder' refers to a television

signal which occupies half the available transponder bandwidth. Two TV signals can be transmitted through a satellite transponder by reducing the deviation and power allocated to each. Half-transponder TV carriers operate typically 4–7 dB below single carrier saturation power.
[See also **saturated output power**.]

HARD DOCK

See **dock**.

HARD VACUUM

Jargon: a 'near-pefect' **vacuum**, as found in **space** (as opposed to a 'partial vacuum' produced in a laboratory).
[See also **vacuum chamber**.]

HARDENING (of SPACECRAFT)

The practice of making the components, and thereby the operation, of a **spacecraft** more resistant to radiation damage, particularly from thermonuclear weapons in a military context, but also from natural radiation (e.g. **cosmic radiation**, **solar wind**, **Van Allen belts**). Solutions vary from simple shielding to improved design of the affected components, which are predominantly semiconductors.
[See also **single-event effect (SEE)**.]

HARDWARE

Any physical equipment. For example: 'flight hardware', any equipment that leaves the surface of the Earth as part of a spacecraft **mission**; **ground support equipment** (GSE); the physical equipment comprising a computer system (cf **software**).

HEAD-END

(i) The central distribution point for a **cable TV** system.

(ii) An alternative term for **head-end unit**.

HEAD-END UNIT

An **amplifier** and/or associated electronics mounted on an earth station **antenna**, at the **prime focus** of a simple **paraboloid** or behind the dish in a Cassegrain or other dual reflector system. An alternative term for LNA, LNB, LNC or 'outdoor unit' [figure L7].
[See also **low-noise amplifier**, **low-noise converter**, **Cassegrain reflector**.]

HEAT EXCHANGER

A device for the transfer of thermal energy from one fluid (liquid or gas) to another without allowing the two to mix—used, for example, in the **rocket engine**.
[See also **cooling system, heat sink, heat pipe, radiator**.]

HEAT PIPE

A device for transferring thermal energy from one point to another. Part of a **thermal control subsystem**.

On board a spacecraft, the heat pipe is used to maintain the temperature of heat-producing components (e.g. **TWT, multiplexer, battery**). Although more complex than a simple **heat sink, heat spreader plate** or **radiator**, the heat pipe is more efficient in transferring heat and removing 'hot spots'.

The principle of the heat pipe is shown in figure H1. The pipe contains a fluid which is vaporised by the applied heat at one end (the evaporator) and condensed at the other end where it relinquishes its heat. The condensed liquid returns to the evaporator through a porous wick or via a system of axial grooves by means of capillary action. This fixed or constant conductance heat pipe (FCHP or CCHP) has a disadvantage in that it transports heat even when it would be useful to retain it—during an **eclipse** for example. The variable conductance heat pipe (VCHP) uses a non-condensable gas (such as nitrogen) to close off part of the condenser when the heat load is reduced, thereby limiting the heat transfer area.
[See also **heat rejection**.]

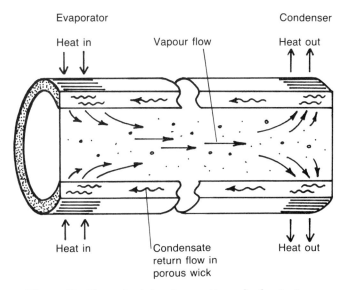

Figure H1 The principle of operation of a **heat pipe**.

HEAT REJECTION

The removal of thermal energy. For a spacecraft this is limited to radiation, since convection and conduction are not possible in a vacuum. Heat loss by evaporation is technically possible but would require excessive amounts of coolant for the long **lifetime**s expected of current spacecraft. Evaporation to the local environment would also degrade the satellite's optical surfaces and **solar cell**s. However, so-called 'closed-loop' evaporation is a process used in the **heat pipe**.

HEAT SHIELD

A coating or surface which offers protection from excessive heat, such as that experienced by a spacecraft on **re-entry**. For example, the 'ablative material' used on the **Apollo** command module was a phenolic epoxy resin, a type of reinforced plastic [see **composites**]. The Apollo heat shield was subjected to temperatures of about 2700 °C.
[See also **ablation, thermal protection system**.]

HEAT SINK

The opposite of a heat source; a passive device for the removal of thermal energy.

The typical spacecraft heat sink is commonly known as a 'heat spreader plate', because its primary function is to spread heat over a wide enough area for it to be radiated away. Although it may resemble a **radiator** and they can be combined, there is a subtle difference: a radiator is primarily a **heat rejection** device, whereas a heat sink temporarily stores and then disperses its heat, tending to reject that heat by conduction (perhaps to a radiator) rather than radiation. It can act, therefore, as a kind of 'thermal capacitor': as the electrical capacitor stores and later releases electrical energy, the heat sink absorbs thermal energy, distributes it throughout its volume and eventually dissipates it into other structures. Heat sinks are, of course, particularly useful for equipment which generates large amounts of heat (e.g. some **radio-frequency** components in a satellite **communications payload**). In general, metals which are good conductors are also good heat sinks, which is why, coupled with its relatively low weight, **aluminium** is widely used.
[See also **thermal control subsystem**.]

HEAT SOAK

The phenomenon whereby the temperature of a **rocket engine** and surrounding or connected components continues to increase after the engine has ceased firing. Sometimes called 'heat soakback'. The effect is due to the 'thermal conduction inertia' of structures which have no

active **cooling system** and is one of the factors which must be considered in a vehicle's thermal design.
[See also **cold soak, thermal control subsystem, heat rejection.**]

HEAT SPREADER PLATE

An item of spacecraft hardware with the properties of a **heat sink**.
[See also **thermal control subsystem.**]

HEATER

A device which uses the heat developed when an electric current passes through an electrical resistance to maintain the temperature of an item of **spacecraft** equipment. Dependent on requirements, a heater may be on continuously, cycled on and off between a maximum and minimum temperature, controlled by a thermostat to maintain a constant temperature, or ground controlled. This flexibility, the power of the heater and its position are all parameters in the thermal design of the spacecraft.
[See also **line heater, thermal control subsystem.**]

HEAVY LIFT LAUNCH VEHICLE (HLLV)

See **heavy lift vehicle (HLV)**.

HEAVY LIFT VEHICLE (HLV)

A **launch vehicle** capable of lifting relatively large and heavy **payloads** into **orbit**. Otherwise called a heavy lift launch vehicle (HLLV). The term is imprecise, but a vehicle that delivers a payload of 4 tonnes or more to **geostationary transfer orbit** could be considered an HLV.

HELIOPAUSE

See **heliosphere**.

HELIOSPHERE

A region of **space** which is influenced by the charged particle flux from the Sun. The boundary between the heliosphere and 'interstellar space' is known as the 'heliopause'. Theory states that outside the heliopause the **solar wind** ceases, and the cosmic-ray intensity is constant and greater than it is inside since it is no longer affected by the solar wind.

The interplanetary **spacecraft** Pioneer 10, which crossed the orbit of Neptune on 13 June 1983, had not reached the heliopause five years later, when it was about 45 AU (**astronomical unit**s) from the Sun, but was expected to reach it before the end of the century.
[See also **cosmic radiation, magnetosphere.**]

HELIOSYNCHRONOUS ORBIT

A type of **polar orbit** in which the satellite's **ground track** remains approximately fixed at the same local time on Earth; an **orbit** which is synchronous with the Sun (also called a 'Sun-synchronous orbit') [figure H2]. This type of orbit has an **altitude** of between about 600 and 800 km and is used for Earth observation or solar study.
[See also **remote sensing, astronomical satellite, low Earth orbit, geostationary orbit**.]

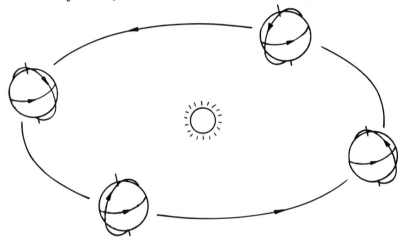

Figure H2 An example of a **heliosynchronous orbit**. The angle between the orbital plane and the Sun direction is constant throughout the year.

HELIX

A type of **slow-wave structure**; a component of a **travelling wave tube**.

HEMISPHERICAL BEAM

A **beam** formed by a **satellite** communications **antenna** which provides coverage of a hemisphere. Although this should mean coverage of half the Earth, the term is invariably used more loosely to include a significant portion of a hemisphere. Moreover, hemispheres are usually divided by lines of longitude rather than by the equator (e.g. a hemispherical beam may provide coverage of North *and* South America).
[See also **global beam, beamwidth, zone beam, spot beam, multiple beam**.]

HERMES

A European manned **space shuttle**, proposed by the French space agency **CNES** in the mid 1980s, which would be built and operated under the auspices of the European Space Agency (**ESA**) and launched by the proposed **Ariane** 5 launch vehicle.

HERTZ (Hz)

The SI unit of **frequency**; the number of cycles or oscillations per second. The most common multiples in use in **satellite communications** are MHz (megahertz, 10^6 Hz) and GHz (gigahertz, 10^9 Hz); kHz (kilohertz, 10^3 Hz) is more common in terrestrial communications and when referring to the **baseband** signal characteristics.

The unit is named after Heinrich Rudolf Hertz, the German physicist, (1857–94), in honour of his experimental confirmation of Maxwell's equations which predicted that an oscillating electrical circuit should radiate an electromagnetic wave. The waves produced became known as 'Hertzian waves'.
[See also **frequency bands**.]

HETERODYNING

Mixing. See **mixer, upconverter, downconverter**.

HETEROGENEOUS PROPELLANT

See **solid propellant**.

HF (HIGH FREQUENCY)

See **frequency bands**.

HIGH-PASS FILTER

A **filter** with a low-**frequency** cut-off which allows only high-frequency signals to pass.
[See also **band-pass filter**.]

HIGH-POWER AMPLIFIER (HPA)

A generic term for a device that amplifies electronic signals to a high power level, usually for transmission over long distances, e.g. between the Earth and a spacecraft in **geostationary orbit**. Sometimes used for a satellite **travelling wave tube amplifier** (TWTA), but more often for an amplifier in an **earth station**.

HIGH-POWER DBS

See **direct broadcasting by satellite**.

HIPHET

See **hiphet thruster**.

HIPHET THRUSTER

A propulsive device utilising electrical heating to increase the **exhaust velocity** of the ordinary **hydrazine thruster**; a type of **reaction control thruster**. Alternatively known as an electrically heated thruster (EHT) or power-augmented hydrazine thruster (PAHT).

Between the **combustion chamber** and the **nozzle**, the combustion products pass through an electrically heated tube, which increases the temperature of the gases, and in turn increases their velocity. Although this improves the **performance** of the standard hydrazine thruster, the hiphet consumes large amounts of electrical power and its use is restricted.

The term is derived from HIgh Performance Electrothermal Hydrazine Thruster (HIPEHT), but it is more often called a 'hiphet' because it is easier to pronounce.

HLV

See **heavy lift vehicle**.

HOHMANN TRANSFER ORBIT

A **trajectory** linking the **orbit**s of two planets, which can be traversed by a spacecraft using a minimum amount of **propellant** (and carrying the maximum **payload**); a 'minimum energy satellite transfer orbit', e.g. **geostationary transfer orbit** (GTO).

The transfer ellipse, named after the German scientist and engineer

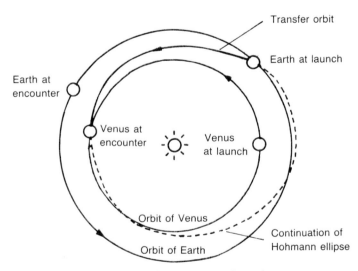

Figure H3 A Hohmann transfer orbit.

Walter Hohmann who published the theory in 1925, is also called a 'Hohmann ellipse' or 'semi-ellipse', since the vehicle follows half of an elliptical path between the two orbits [figure H3]. The path is tangential to both orbits and has the Sun or a **planetary body** at one focus. For interplanetary trajectories, if the spacecraft is launched from Earth to a planet further from the Sun, **launch** is at **perihelion** and **encounter** is at **aphelion**; for planets closer to the Sun, launch is at aphelion, encounter at perihelion. The main drawback of the 'Hohmann transfer' is the relatively long journey time: from Earth to Mars, for example, would take about 260 days; Saturn would be six years away. However, many planetary exploration **probe**s have used the trajectory for at least the first leg of their journey [see **gravitational boost**].
[See also **escape velocity**, **launch window**.]

HOLD (in COUNTDOWN)

A period during which the **countdown** to a **launch** is halted, due either to an unforeseen problem or, with more complex vehicles, as a pre-planned measure to allow time for any procedures which have fallen behind to 'catch up', and for final checks (e.g. on the weather) to be made. Such contingency measures are known as 'built-in holds'.
[See also **recycle countdown**.]

HOLD-DOWN ARM

One of a set of devices which holds a **launch vehicle** on the **launch pad** until its **thrust** has increased to a level sufficient to ensure a successful **lift-off**. Also termed 'launcher release gear'. Typically, the arms are counterbalanced to ensure that they rotate away from the launch vehicle 'under gravity', as soon as the pyrotechnic release charges are fired.
[See also **pyrotechnic cable-cutter**.]

HOMOGENEOUS PROPELLANT

See **solid propellant**.

HONEYCOMB PANEL

A type of structural panel commonly used on spacecraft to reduce the overall mass. It is generally a 'sandwich' of two **face-skin**s bonded to a cellular 'honeycombed' core with epoxy-film adhesive [figure C1]. The core is typically an **aluminium** alloy, the face-skins aluminium or **titanium** alloys, or some form of **carbon composite**.

Typical applications: spacecraft body panels, **antenna** reflector cores.
[See also **materials**.]

HORIZON SCANNER

See **earth sensor**.

HORIZON SENSOR

See **earth sensor**.

HORIZONTAL POLARISATION

See **polarisation**.

HORN

See **feedhorn**.

HORN ANTENNA

See **antenna**.

HORUS

See **Sanger**.

HOTOL

A **single-stage-to-orbit** unmanned **aerospace vehicle** proposed by Britain as a next-generation satellite launcher. HOTOL is an acronym for HOrizontal Take Off and Landing, due to its ability to take off and land on a runway like a conventional aircraft. Its **propulsion system** comprises an air-breathing engine which can also be used as a **rocket engine**, using **liquid oxygen** and **liquid hydrogen** propellants, once above the atmosphere.
[See also **Sanger**.]

HOUR-ANGLE

The angle between the observer's meridian and the meridian of the subject under observation, measured in the equatorial or geographical system of celestial coordinates. Hour-angle is measured towards the west, from 0 to 360 degrees; in astronomy, from where the term originates, it is usually expressed in units of time, where 15° equals one hour. See also **declination**: knowledge of a **satellite**'s declination and hour-angle allow an **earth station** to be aligned with it. An alternative 'horizon system' uses **altitude** (or **elevation**) and **azimuth**.

HOUSEKEEPING

The function of a number of **subsystem**s, on both manned and unmanned **spacecraft**, involved in maintaining the spacecraft in an **operational** condition (e.g. **telemetry**, a certain degree of **attitude control** and thermal control and, in a **manned spacecraft, life support**).

HPA

See **high-power amplifier**.

HPBW

See **half-power beamwidth**.

HTPB

An acronym for the **solid propellant** hydroxyl terminated polybutadiene. See **polybutadiene**.

HUBBLE SPACE TELESCOPE

See **space telescope**.

HYBRID

A **microwave** device used to connect more than one **input** path to more than one **output** path, in a satellite **transponder** for example. A hybrid is generally located in a similar manner to a switch: the difference is that a switch position has to be 'selected' (by ground **command**); a hybrid is a **passive** device which links several input and output paths 'simultaneously'.

HYBRID ROCKET

A rocket which uses one **propellant** in the solid phase and another in the liquid phase [see **solid propellant, liquid propellant**]. Although hybrids have not been used extensively, the most common combination is a solid **fuel** and a liquid **oxidiser** (e.g. **polybutadiene** and **liquid oxygen**), the latter being sprayed onto the former to initiate **combustion**. The hybrid generally has a higher **performance** than a solid motor and is less complex than a liquid engine, although it retains the cut-off/restart capability of the latter [see **engine re-start**].

HYDRAZINE (N$_2$H$_4$)

A storable liquid **fuel** used in **rocket engines**. See **liquid propellant**. [See also **hydrazine thruster, hiphet thruster, monopropellant, MMH, UDMH**.]

HYDRAZINE THRUSTER

A propulsive device utilising the chemical compound **hydrazine** (N_2H_4) as a **propellant**; a type of **reaction control thruster**. The propellant is stored as a liquid and usually fed to the thruster by a **pressurant** such as gaseous nitrogen or helium. It is decomposed to ammonia, nitrogen and hydrogen with the aid of a catalyst in a **combustion chamber** and the hot gases are ejected from the **nozzle** in a tiny jet (hence the colloquial term for reaction thrusters, 'gas jets'). The hydrazine thruster can deliver its **thrust** in short or long bursts, or in pulses as short as a few milliseconds.

[See also **pulsed thrust, hiphet thruster, cold-gas thruster, bipropellant thruster, gas generator**.]

HYDROGEN PEROXIDE (H_2O_2)

A storable liquid **oxidiser** used in **rocket engines**. See **liquid propellant**.

HYPERGOLIC

An attribute of a **bipropellant** combination (**fuel** and **oxidiser**) which ignites spontaneously when mixed (e.g. monomethyl hydrazine (**MMH**) (fuel) and **nitrogen tetroxide** (oxidiser)). Propulsion systems using hypergolic propellants are simplified by the lack of an **ignition system**. [See also **pyrophoric fuel**.]

HYPERSONIC FLOW

See **mach number**.

HYPOXIA

A deficiency in the amount of oxygen in the blood and body tissues. Hypoxia becomes evident at an altitude of about 3.5 km above the surface of the Earth, when the alveoli of the lungs experience difficulty in absorbing oxygen at the ambient pressure. A complete lack of oxygen, experienced at about 16 km, is called 'anoxia': this occurs when the ambient pressure drops to about 11.6 kPa (1.7 psi or 87 mmHg), the pressure maintained within the alveoli, at which point there can be no transfer of oxygen across the alveoli membrane.

[See also **decompression sickness, ebullism**.]

I

IAF

An acronym for the International Astronautical Federation, a body founded in 1950 to develop **astronautics** for peaceful purposes by means of public education, the dissemination of technical information, and the organisation of congresses and scientific meetings. In 1960 it created the cooperative but autonomous International Academy of Astronautics (IAA) and International Institute of Space Law (IISL).

ICBM

An acronynm for InterContinental Ballistic Missile. Many early American ICBM boosters were adapted for use in the space programme (e.g. the **Atlas** booster used to launch the **Mercury** spacecraft and the **Titan** II used for **Gemini**). In the late 1980s the Atlas and the Titan III became commercial launchers of satellite payloads.
[See also **ballistic trajectory**.]

IF

See **intermediate frequency**.

IF AMPLIFIER

A device which amplifies a signal at an **intermediate frequency** (IF).

IFRB

An acronym for International Frequency Registration Board, the body within the International Telecommunications Union (**ITU**) to which all applications to use a part of the **radio spectrum** are submitted (e.g. for **communications satellite** services, etc). The IFRB ensures that there is no conflict between services using the limited resource of radio frequencies and **orbital position**s which might lead to **interference**, and records assignments of these resources.

IGLOO

See **Spacelab**.

IGNITER

A device in a **rocket engine** or **rocket motor** which initiates **propellant** combustion. See **ignition system**.
[See also **percussion primer**.]

IGNITION

The act or process of initiating **combustion** by igniting the **propellant**(s) in a **rocket motor** or **rocket engine**.
[See also **ignition system**.]

IGNITION SYSTEM

The part of a propulsion system which initiates **combustion** in a **rocket engine** or **rocket motor**. Ignition systems may utilise **pyrotechnics**, spark-ignition or electrically heated 'hot-wire' devices. For the typical **solid propellant** motor, a **power supply** sends a current pulse to a pyrotechnic cartridge which ignites a small sample of the propellant in a steel or glass-fibre housing. This produces a controlled amount of hot gas which, in a similar way to a detonator in an explosive device, ignites the main motor.

In the case of a **liquid propellant** engine, the propellants (**fuel** and **oxidiser**) are supplied either by a pressurising gas (**pressurant**) or by mechanical **turbopump**s and are injected into a **combustion chamber**, atomised and mixed. Initially a small amount of fuel meets the oxidiser and a 'pilot flame', produced by the ignition system, heats the liquids to vaporisation point. If the propellants are not self-igniting (**hypergolic**), one of the above devices is used to produce the heat required for evaporation, ignition and final combustion (if they are hypergolic, there is no need for an ignition system). After ignition the supply valves are opened fully and combustion temperatures and pressures rapidly increase to produce a high-velocity exhaust.
[See also **nozzle, exhaust velocity, percussion primer**.]

IGY

An acronym for International Geophysical Year, an international scientific programme jointly conducted by 54 nations between July 1957 and December 1958. Its principal task was the comprehensive study of the influence of solar activity on the various phenomena observed in the atmosphere, **ionosphere** and near-**space**. Thus the time selected for the IGY corresponded to a period of maximum solar activity. Particularly useful were the first artificial **satellites** carrying scientific instruments: analysis of part of the **data** returned led to the identification of the **Van Allen belts**. IGY studies continued into 1958–59 were referred to as the International Geophysical Cooperation Year (IGCY). A similar programme to the IGY, known as the International Quiet Sun Year (IGSY), was carried out during the 1964–65 solar-minimum period.

IMAGE

A reproduction of a physical object or scene formed by a lens, mirror, or some other focusing system, projected onto an imaging device and displayed using a medium such as photographic film or **television**. The 'reproduction' is referred to as an image at all stages of the process: it may be the image formed by a **telescope** sensitive to any **wavelength** of radiation (x-ray, radio, etc, as well as optical); it may be the image on the photosensitive imaging tube of an electronic camera or the retina of the eye; or it may be the image on a screen, a piece of paper, or any other 'output device'. The process of obtaining 'an image' is generally known as 'imaging'.
[See also **imaging team, pixel.**]

IMAGE INTENSIFICATION

The process concerned with increasing the 'clarity' of an **image**. It is not so much that information is added to the image, more that the information it already contains is made more readily accessible—increasing the image's intelligibility. This can be seen at its simplest as an increase in contrast or an enhancement of **signal-to-noise ratio**. An advanced form of the technique is used to improve the **raw data** received from spacecraft imaging systems to produce more intelligible images—this is often referred to as computer enhancement.
[See also **resolution.**]

IMAGING

See **image**.

IMAGING TEAM

The collective term commonly used for a group of scientists and engineers concerned with the reception and interpretation of **images** received from a planetary exploration **probe** or other similar spacecraft.

IMPULSE

A product of the force (or **thrust**) acting upon a body and the time for which it acts (measured in N s). Used in connection with **propulsion systems**: e.g. 'the impulse of a **rocket engine**'.
[See also **specific impulse**.]

IMUX

A colloquial abbreviation for input multiplexer. See **multiplexer**.

IN-ORBIT SPARE

A **satellite** in **orbit** which is available to replace a defective satellite or enhance an existing satellite system. It usually remains inactive until needed, but it may be incorporated into the system on a part-time or part-**capacity** basis. The provision of an in-orbit spare can change the status of a satellite system from **pre-operational** to **operational**, since it offers a guarantee that the service will continue if the **prime** satellite fails.
[See also **ground spare**.]

INCLINATION

See **orbital inclination**.
[See also **declination, elevation (angle)**.]

INDOOR UNIT

A colloquial term for the package of DBS **receiver** electronics connected between the **head-end** or **outdoor unit** and the **television** set in a direct broadcast system.
[See also **direct broadcasting by satellite**.]

INERTIA WHEEL

See **momentum wheel** and **reaction wheel**.

INERTIAL FRAME

In physics, an unaccelerated reference frame (e.g. the 'fixed stars'). The Foucault pendulum experiment, first performed publicly in 1851 using a mass on a wire nearly 70 m long, demonstrates that the Earth is a rotating, non-inertial frame: the plane of motion of the pendulum remains fixed in an inertial frame (i.e. fixed relative to the stars) while the Earth rotates beneath it.
[See also **inertial platform, frame of reference**.]

INERTIAL GUIDANCE

See **inertial platform**.

INERTIAL PLATFORM

A device which uses a set of **gyroscopes** to define a **frame of reference** in a moving vehicle (e.g. a **launch vehicle**), with respect to which the vehicle's **attitude** can be controlled; a platform stable with respect to inertial space [see **inertial frame**].

An inertial platform typically comprises three gyros and three accelerometers mounted on **gimbals** which isolate them from the vehicle. The gyros are set spinning with their axes in the three orthogonal directions which, however the vehicle moves, will remain fixed in inertial space. Any movement of the gimbal frames around the gyros will be detected and the signal thus generated used to drive the gimbal motors to hold the gyro cluster in a fixed orientation. The stable gyro platform provides the accelerometers with a reference. Any acceleration imposed upon the platform is detected by the accelerometers, the outputs from which are processed to calculate the vehicle's velocity and position. The actual **trajectory** is compared with a theoretical trajectory and the vehicle **autopilot** corrects any errors. This is known as inertial guidance or navigation.

INERTIAL UPPER STAGE

A particular type of **upper stage**, used on the American **Space Shuttle** and **Titan** expendable launch vehicle (ELV), which combines the functions of **perigee kick motor** and **apogee kick motor** [see figure T1]. The IUS comprises two **solid propellant** stages, both of which separate from the **spacecraft** after firing (i.e. a spacecraft launched using an IUS has no integral apogee motor). The IUS also carries its own guidance and control **avionics**: it is **three-axis stabilised** and inertially navigated [see **inertial platform**]. Its **payload** capability when used on a Titan 34D is 1870 kg to **geostationary orbit** (GEO); it is 2270 kg when used with the Shuttle. Its first flights on Titan and Shuttle were in October 1982 and April 1983, respectively.
[See also **payload assist module (PAM)**, **Centaur G**.]

INGRESS

An entrance; the act of entering. A word commonly used as a modifier (e.g. 'ingress hatch', 'ingress port'). The opposite of **egress**.

INHIBITOR

See **flame inhibitor**.

INJECTION

(i) The process of putting a spacecraft in an **orbit**—orbital injection or insertion.

(ii) The time following **launch** when a vehicle is following a **ballistic trajectory**.

(iii) The introduction of **propellant**, etc, into an engine, **combustion chamber**, etc.

INMARSAT

The International Maritime Satellite Organisation, established in July 1979 (following the adoption of its convention in September 1976) to develop and promote a global maritime **satellite communications** system. Inmarsat began operations in February 1982 when it took control of satellites previously operated by 'Marisat' (an American joint venture) and expanded the network by leasing satellites from the European Space Agency (**ESA**) and **Intelsat**. As of November 1989 Inmarsat had 58 signatories (member countries).

INPUT

(i) The entry point or path through which information or energy is applied to a device or system (refers to the hardware).

(ii) The information or energy (a **signal, carrier, RF wave**, etc) fed into an electronic or **microwave** circuit or component.
[See also **input section, input filter, data**.]

INPUT FILTER

A **filter** at the end of a **receive chain** in a communications **transponder** which confines the signal to its appointed **bandwidth**. This type of device is known generically as a **band-pass filter** and is often installed as part of an input **multiplexer**.
[See also **output filter**.]

INPUT SECTION

The section of a spacecraft **communications payload** from the point where the input signal enters from the receive antenna to the point prior to its entry to the **travelling wave tube amplifier**. The division of payload components into an input section and an **output section** is an alternative to their classification as part of a **receive chain** or a **transmit chain** [see figure C2].

INSERTION

See **injection**.

INSTRUMENT UNIT

The part of a **launch vehicle** containing the guidance and control equipment, **telemetry** transmitter and other electronic devices [see figure I1]. Also called instrument bay, vehicle equipment bay, etc. On contemporary three-**stage** vehicles, for example, the instrument unit is an integral part of the third stage, the last stage to be ignited. [See also **autopilot**.]

Figure I1 Giotto mounted on its **Ariane** 1 launch vehicle. Note the launch vehicle **instrument unit** on the top of the third stage. [ESA]

INSULATION

See **thermal insulation**.

INTEGRATED SERVICES DIGITAL NETWORK (ISDN)

An international network (operated by the **PTTs**) for the provision of **telecommunications** services. The internationally agreed definition of ISDN is 'a network evolved from the telephony IDN (integrated digital network) that provides end-to-end digital connectivity to support a wide range of services, including voice and non-voice services, to which the users have access by a limited set of standard multipurpose customer interfaces'. From the customer's point of view, ISDN offers access to a modern, digital network (System X in the UK) capable of handling both data and telephony transmissions.

INTEGRATION

See **spacecraft integration, payload integration**.

INTELSAT

The International Telecommunications Satellite Organisation, created in August 1964 (originally as 'The Provisional Committee for Satellite Telecommunications') to develop the first worldwide satellite communications network. It offers **telecommunications** services to all parts of the world except the polar regions, using **satellite**s in **geostationary orbit**.
[See also **Intelsat standard earth stations, Intersputnik, Inmarsat**.]

INTELSAT STANDARD EARTH STATIONS

Earth stations authorised by **Intelsat** to operate within the Intelsat system. There are three standards for international **gateway station**s:
   standard A: originally antennas 30 m in diameter or larger, operating at C-band (now 18 m due to increased satellite power);
   standard B: antennas 11 m in diameter or larger, operating at C-band (for lower **traffic** demands than Standard A);
   standard C: originally antennas 14–19 m in diameter or larger, operating at Ku-band (now 13–15 m).
   There are also a number of standards for smaller earth stations:
   standard D1: 4.5 m diameter at C-band (for use within the 'Vista' low-density **SCPC** telephony service);
   standard D2: 11 m diameter at C-band (for use within the 'Vista' low-density SCPC telephony service;
   standards E and F: 3.5–9.0 m diameter at Ku- and C-band respectively (for use within the international 'Intelsat Business Service' (IBS));
   standard G: wide range of earth terminal sizes down to 0.8 m 'microterminals' (for use with the 'Intelnet' **data** distribution service).
   standard Z: 3–18 m diameter at C- and Ku-band (for use with Intelsat domestic lease services).
[See also **frequency bands**.]

INTERCONTINENTAL BALLISTIC MISSILE

See **ICBM**.

INTERCOSMOS

A system for collaboration in **space science** and **astronautics** between the Soviet Union and other nations which has developed since the launch of the first **Sputnik** satellites in 1957. Contemporary agreements under the Intercosmos banner include joint Soviet manned spaceflights with members of other Eastern Bloc countries. Also included are French **cosmonaut**s, following a Franco–Soviet agreement on cooperation in space science signed in June 1966.

INTERFACE FILLER

A material placed between two surfaces to improve the thermal conductivity between them. Also called 'thermal grease'.

INTERFERENCE

In **telecommunications**, a term applied to the undesirable superposition of two or more **radio-frequency** (RF) signals in a **receiver**. In a satellite **communications system**, there are three main factors used to reduce interference: the allocation of different frequencies (**frequency separation**); the use of opposite **polarisation**s; and the angular separation between satellites in different **orbital positions** on the **uplink** and that between **earth station**s in different geographical areas on the **downlink**.
[See also **multipath, frequency coordination, interference analysis, protection ratio.**]

INTERFERENCE ANALYSIS

A comparison of radio communication parameters, such as **carrier** powers at the **input** to a **receiver**, which indicates whether or not a particular service will suffer radio **interference** from other services. If the power of the wanted carrier is sufficiently greater than that of the unwanted carriers, there will be no interference.
[See also **isolation.**]

INTERMEDIATE FREQUENCY (IF)

A **frequency** used in a communications **transponder**, or other receive/transmit device, between the input and the output. In a satellite communications link the **uplink** and **downlink** frequencies must be different to obviate interference: the uplink frequency is the higher of the two, since the greater (**atmospheric attenuation**) losses at higher frequencies can be more readily overcome by higher power earth station transmitters (cf increasing the spacecraft power). In the so-called 'double-conversion' transponder, the intermediate frequency is the **output** of the **downconverter** and the **input** to the **upconverter**. A typical example for C-band might be: uplink 6 GHz; IF 750 MHz; downlink 4 GHz.

It is possible to convert directly from the received uplink frequency to the transmitted downlink frequency (so-called 'single conversion'), but a signal at IF is easier to handle than the high-power, high-frequency **carrier** wave required for long-distance communication. The technology for amplification, filtering and switching is far simpler at the lower frequencies, and the hardware is easier to produce and therefore less expensive. In addition, the lower frequencies can be conducted by **coaxial cable** which, since it is lighter than **waveguide**, is an important factor for the **mass-limited** spacecraft **payload**.

INTERNATIONAL ASTRONAUTICAL FEDERATION

See **IAF**.

INTERNATIONAL FREQUENCY REGISTRATION BOARD

See **IFRB**.

INTERNATIONAL MARITIME SATELLITE ORGANISATION

See **Inmarsat**.

INTERNATIONAL TELECOMMUNICATIONS SATELLITE ORGANISATION

See **Intelsat**.

INTERNATIONAL TELECOMMUNICATIONS UNION

See **ITU**.

INTER-ORBIT LINK (IOL)

A **communications** link between spacecraft in different **orbits**, e.g. between a communications or data-relay satellite in **geostationary orbit** and a **remote sensing** spacecraft or **space station** in **low Earth orbit** [see figure I2].
[See also **inter-satellite link**.]

INTER-SATELLITE LINK (ISL)

A **communications** link between spacecraft in **orbit** [see figure I2]. ISLs utilise different radio frequencies to the fixed and mobile services and may use laser light. The link obviates the need for double hopping, the transmission of a signal up to a satellite and down to an **earth station** twice in succession in order to span a large distance around the Earth [see **double hop**].

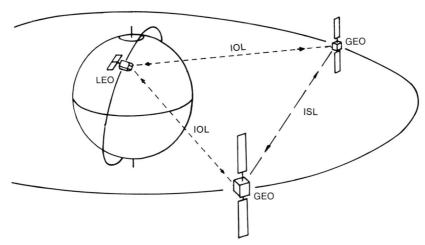

Figure I2 Inter-satellite links between spacecraft in **geostationary orbit** and **low Earth orbit**.

Alternatively, ISLs can be used to link two or more small satellites in relatively closely spaced **orbital positions** (a 'local ISL'), so that the cluster acts as a large-capacity satellite. ISLs can also reduce the number of earth stations needed, by allowing the user to point a single earth station at any one satellite and communicate with the others through the ISLs.

The term inter-satellite link is generally reserved for spacecraft in the same orbit, usually **geostationary orbit**. When the link is between two spacecraft in different orbits, say geostationary orbit and a **low Earth orbit**, it is termed an **inter-orbit link** (IOL).
[See also **fixed-satellite service, mobile-satellite service**.]

INTER-SATELLITE SERVICE (ISS)

A **satellite communications** system providing links between spacecraft orbiting the Earth—**inter-satellite links** (ISLs).
[See also **fixed-satellite service (FSS)**.]

INTERSPUTNIK

The Eastern Bloc telecommunications satellite organisation established in 1968.
[See also **Intelsat**.]

INTERSTAGE

A section of a **launch vehicle** between two **stages**, which surrounds the

engine(s) of the uppermost stage. It may be an open latticework structure, or 'faired in' with the rest of the vehicle to provide aerodynamic continuity between stages. It is usually jettisoned during **staging**, before **ignition** of the upper stage.
[See also **fairing**.]

INTERTANK STRUCTURE

In a **launch vehicle**, a supporting structure between two **propellant** tanks which enhances the strength of the **stage** and provides continuity of the **load path**. It also provides a protected site for instruments, etc.

INTRA-VEHICULAR ACTIVITY (IVA)

Any activity which takes place inside a **spacecraft**; the opposite of **extra-vehicular activity** (EVA).

INTRA-VEHICULAR PRESSURE GARMENT

See **spacesuit**.

INVAR

An alloy composed of iron, nickel and carbon which has a low coefficient of expansion; a tradename: an abbreviation of 'invariable'. Used to manufacutre **radio-frequency** components (e.g. **filter**s, input and output **multiplexer**s), etc.

IOL

See **inter-orbit link**.

ION ENGINE

A propulsive device in which electrical energy is used to ionise a gaseous **propellant** and an electrostatic field accelerates the positive ions to produce **thrust**. Also called an ion thruster.

In one device (the Kaufman electron bombardment ion thruster), electrons bombard the atoms of the propellant vapour, forming a cloud of positive ions. This cloud drifts towards a pair of perforated electrode plates (known as the 'screen grid' and 'accelerator grid') which together form the 'exit port' of the discharge chamber. The ions are accelerated by the high potential gradient across the grids and ejected as a high-velocity exhaust. As an alternative to electron bombardment, ionisation can be induced in a quartz discharge chamber by a **radio-frequency** (typically 1 MHz) generator coil. This type of device is generally known as a radio-frequency ionisation thruster (RIT).

Ion engines have a **specific impulse** (I_{sp}) an order of magnitude greater than those available from chemical propulsion systems. For example, RIT devices using xenon as the propellant have demonstrated an I_{sp} of around 3500 s with a thrust of about 10 mN. Although the low thrusts available from ion engines have usually suggested applications on interplanetary spacecraft, which could benefit from low thrust over a long period, the engine has potential applications on satellites, e.g. north–south **station keeping** and fine **attitude control**, **orbital relocation** and transfer to a **'graveyard orbit'** for satellites in **geostationary orbit**, and **drag compensation** for those in **low Earth orbit**.

[See also **electric propulsion**.]

ION PROPULSION

See **electric propulsion**.

ION THRUSTER

See **ion engine**.

IONISATION

The formation of ions. In chemistry, ions can be positively or negatively charged (cations or anions), but to the physicist they are generally positively charged atoms. Ionisation may be the result of chemical reaction, heat, electrical discharge or radiation bombardment, all of which cause **electron**s to be liberated from or 'stripped off' the neutral atom, leaving a net positive charge.

[See also **ionosphere, plasma sheath**.]

IONOSPHERE

A region of the Earth's atmosphere, in which ionising radiation (chiefly ultraviolet and x-ray emission from the Sun) causes the liberation of **electron**s [see **ionisation**]. The resultant 'ionised gas' of ions and electrons is called a plasma.

The depth of the ionosphere varies with the extraterrestrial particle flux, and therefore with the local time of day: during the daylight hours its lower reaches extend to an altitude of about 50 km, but at night it lifts to about 80 km. The top of the ionosphere is about 1000 km above the surface, but this depends on the value of plasma density used in the definition since the ionosphere can be thought of as 'thinning into' and contiguous with the interplanetary plasma.

Since a plasma contains 'free electrons', capable of independent movement relative to the surrounding ions, the layers of ionised gas have properties similar to those of a metallic conductor, in that they can reflect radio waves. It was this property which led to the experimental determination of the ionosphere's structure, by the English physicist Sir Edward Victor Appleton (1892–1965), in the 1920s. This knowledge has since been usefully exploited for certain types of terrestrial communication.

Especially at times of high particle flux, for instance due to a **solar flare**, particles entering the ionosphere lead to the displays known as the Aurora Borealis and Aurora Australis.
[See also **ionisation, magnetosphere**.]

I^2R LOSS

The loss of **power** in a **transmission line** due to heating. The mathematical symbology derives from Joule's law for the production of heat in a conductor, otherwise known as 'the Joule effect' or 'Joule heating'. The effect is particularly important for long transmission lines: e.g. the **waveguide** runs in large **earth station**s, between the **feedhorn** and the **high-power amplifier** (HPA), are generally kept to a minimum by mounting the feedhorn and HPA close together.

ISDN

See **integrated services digital network**.

ISL

See **inter-satellite link**.

ISRO

An acronym for the Indian Space Research Organisation.

ISS

See **inter-satellite service**.

ISOCHRONOUS

In **telecommunications**, a property of a transmission where the signals are of equal duration and are sent in a continuous sequence.

ISOFLUX CONTOURS

Contours on a satellite antenna **footprint** which represent the same level of incident flux, or **power flux density** (e.g. 'the – 3 dB contour', which is '3 dB down' from the **peak**) [see figure F1].
[See also **antenna, beamwidth, coverage area**.]

ISOLATION

In **telecommunications**, a measure of separation, in terms of **RF power**, between the **carrier** waves of two potentially interfering communications systems. The isolation, measured in dB, may be afforded by transmitting on different **polarisation**s or in different directions (e.g. into separate satellite antenna **beam**s).

In an **earth station**, isolation entails preventing high transmission powers from entering the receiver, which is achieved largely by filtering. [See also **interference analysis, channel isolation**.]

ISOTROPIC

Having uniform physical properties in all directions.

ISOTROPIC ANTENNA

A hypothetical **omnidirectional** point-source **antenna** used as a reference for **antenna gain** measurements. Such an antenna radiates equally in all directions and can therefore be defined as having 'no gain' in any direction, i.e. no **directivity**. (To be precise, 'no gain' here means a gain value of '1' numerically, or a gain of 0 dB). [See also EIRP **(equivalent isotropic radiated power)**.]

ISY

An acronym for International Space Year, a programme to promote multinational cooperation in space, modelled on the International Geophysical Year [see **IGY**], and proposed for 1992 (selected in honour of the 500th anniversary of Columbus' discovery of America and the 75th anniversary of the Russian revolution).

ITU

An acronym for International Telecommunications Union, a body responsible for the planning and regulation of international **telecommunications** services [see **fixed-satellite service (FSS)**, **broadcasting-satellite service (BSS)**]. For administrative purposes, the ITU divides the world into three regions: Region 1 includes Europe, Africa, the USSR and Mongolia; Region 2 the Americas and Greenland; and Region 3 Asia, Australasia and the Pacific. [See also **CCIR, CCITT, IFRB, WARC, RARC**.]

IUS

See **inertial upper stage**.

IVA

See **intra-vehicular activity**.

J

JET PROPULSION LABORATORY

See **NASA**.

JETTISON

To cast off or release. The word is typically used for items which are being discarded, no longer of any use (e.g. spent rocket **stages**, **heat shield**s, aerodynamic **fairing**s, rubbish containers, etc); in cases where the item is not being discarded (e.g. a **satellite** released from the American Space Shuttle **orbiter**), it is more usual to use the terms '**launch**', 'release', etc.

JIUQUAN

A Chinese **launch site**, formerly known in the West as Shuang Chen Tse (or Shuang-ch'eng-tzu (alternative spelling): 'East Wind'), at approximately 41°N, 100°E on the edge of the Gobi Desert. Until 1987, most of China's satellites were launched from Jiuquan and it is still used for recoverable **reconnaissance satellite**s and Earth resources satellites, but launches into **geostationary orbit** are now made from the **Xichang** site.
[See also **Taiyuan**.]

JOHNSON SPACE CENTER (JSC)

See **NASA**.

JOULE–THOMSON EFFECT

The decrease in temperature when a gas expands; a phenomenon which proved useful in the production of **cryogenic propellant**, amongst other things. (Note: under certain conditions a reverse effect can occur, i.e. temperature increases). Also called the Joule–Kelvin effect (since the Scottish mathematician and physicist, William Thomson (1824–1907) became Lord Kelvin in 1892).

JPL

An acronym for Jet Propulsion Laboratory. See **NASA**.

JUPITER

An American single-**stage** liquid propellant intermediate range ballistic missile (IRBM) based on the US Army **Redstone** and developed in the late 1950s (**propellant: liquid oxygen/kerosene**). Used in May 1959 to launch a **capsule** containing two monkeys (named Able and Baker) on a **ballistic trajectory** to test, amongst other things, their reaction to the effects of high acceleration and **weightlessness**. No longer **operational**. [See also **Thor**.]

K

K-BAND

18–26.5 GHz—see **frequency bands**.

Ka-BAND

26.5–40 GHz—see **frequency bands**.

Ku-BAND

12.5–18 GHz—see **frequency bands**.

KELVIN (K)

The SI unit of temperature, named after the Scottish mathematician and physicist William Thomson, Baron Kelvin of Largs (1824–1907); the unit of **absolute temperature**. Thomson suggested that -273 °C should be considered the 'absolute zero' of temperature, at which the kinetic energy of molecules would be zero, and that the degrees on the 'absolute scale' should be the same as those on the Celsius scale (devised in 1742 by the Swedish astronomer Anders Celsius (1701–1744)). The modern figure for absolute zero is about -273.15 °C. Although it is more or less obligatory to use the (°) symbol for °C, it is more common, by convention, to write 273 K rather than 273 °K. [See also **rankine (°R)**.]

KENNEDY SPACE CENTER (KSC)

NASA's main **launch** facility (located at approximately 28.5°N, 81°W), known more fully as the John F Kennedy Space Center, after the past president of the USA; also known as 'Cape Canaveral', after the part of Florida upon which it is built. The adjacent 'Eastern Test Range' (ETR), known prior to May 1964 as the 'Atlantic Missile Range', is the site of

US Air Force launch facilities. When KSC was established in July 1962 the land was renamed 'Cape Kennedy', but public confusion between NASA and USAF facilities and local opposition to the name change led to the reinstatement of the original title in 1973–74, when the USAF facility became 'Cape Canaveral Air Force Station'. KSC also manages NASA launches conducted at the 'Western Test Range' in California [see **Vandenberg Air Force Base (VAFB)**].
[See also **vehicle assembly building (VAB)**, **orbiter processing facility (OPF)**, **Shuttle landing facility (SLF)**, **launch control center (LCC)**, **launch complex.**]

KEROSENE

A storable liquid **fuel** used in **rocket engine**s. Also referred to as RP-1. See **liquid propellant**.

KEVLAR COMPOSITE

A material composed of a polymer matrix reinforced by threads or fibres of Kevlar (a form of the organic polymer polyparabenzamide; a proprietary material developed by Du Pont). Otherwise known as Kevlar-reinforced plastic (KRP). Kevlar exhibits a typical density only 55% that of aluminium with the additional advantage of improved impact resistance.

Typical applications: spacecraft **face-skin**s, **antenna** reflector surfaces, filament-wound motor cases and structural components.
[See also **composites**, **carbon composite**, **materials.**]

KINETIC HEATING

See **aerodynamic heating**.

KLYSTRON

A type of 'electron tube' used as a **high- (RF) power amplifier** in radar systems and satellite **earth station**s. Another example of an electron tube is the **travelling wave tube**, used in both earth stations and satellites.

KOUROU

The location of the **Guiana Space Centre**, the **launch site** of the European **Ariane** launch vehicle.

KRP

An acronym for Kevlar-reinforced plastic. See **Kevlar composite**.

KSC

See **Kennedy Space Center**.

KVANT

See **Mir**.

L

L-BAND

1–2 GHz—see **frequency bands**.

LAE

See **liquid apogee engine**.

LAGRANGE POINT

One of a number of points in space where gravitational forces are balanced so that a body at that point remains nominally stationary. Also called Lagrangian point or libration point.

The existence of these points was calculated by the Italian–French astronomer and mathematician Joseph Louis Lagrange (1736–1813), who predicted that three astronomical bodies could move in a stable configuration, with one at each of the vertices of an equilateral triangle, if one of the bodies was small. The discovery of Jupiter's 'Trojan asteroids' in the early 20th century proved the theory: they orbit in two groups 60° ahead and 60° behind the planet, in its orbit about the Sun. Similar locations in Earth orbit have been suggested for space stations and 'space colonies'.

LANDSAT

A series of American Earth resources satellites [see figure L1].
[See also **remote sensing**.]

LANGLEY RESEARCH CENTER

See **NASA**.

LASER

An acronym for light amplification by stimulated emission of radiation.
A device for producing a narrow, monochromatic beam of coherent
light (i.e. of a single **frequency** and constant relative phase). The laser
was a later result of the development of the first **maser (microwave**
amplification, etc), in 1953, by Charles Hard Townes, an American
physicist. In 1958 Townes and his brother-in-law published a paper on
the possibility of an 'optical maser', for the amplification of infrared or
visible light instead of microwaves. The first practical laser was
constructed by Theodore Harold Maiman, another American physicist,
in 1960.

In terrestrial **communications**, the laser is used with **optical fibre**
links; in space, one of the major potential uses is in **inter-satellite link**s.
It can also be used for the measurement of distance—see **laser ranging
retro-reflector**.

Figure L1 The **Landsat**-D **remote sensing** satellite in a **cleanroom**.
Note the communications **antenna** on top and **solar array** panels
folded around the sides. The satellite is covered in **thermal
insulation** foil.

LASER GYRO

A type of **gyroscope** which uses a **laser**, instead of a spinning mass, to detect rotation (which is used as an input to a **launch vehicle** guidance system). Also known as a 'laser ring gyro' due to the typical arrangement of three mirrors in a triangle, with one laser beam propagating in a clockwise direction and another travelling anticlockwise. The operation of the laser requires a standing wave to be maintained between the mirrors, which implies the existence of a node in the wave pattern at each of the mirrors. However, if the device rotates, a particular wavefront on one beam has a little further to go before it reaches a mirror (conversely the other beam reaches the mirror sooner). Automatic adjustments mean that the wavelength of the first beam is increased to maintain the standing wave, and that of the other beam is reduced. The light from the two beams is incident on a photodetector, where the frequency difference generates a beat frequency. This frequency is proportional to the rotation rate of the device. A complete **inertial platform** uses three such laser rings, one for each axis, 'strapped down' to the vehicle rather than mounted in a set of **gimbal**s [See **strap-down gyro system**].

LASER RANGING RETRO-REFLECTOR

A type of reflector used in the measurement of distance which returns incident **laser** radiation to the point of its origin, i.e. it ensures that the outgoing rays are parallel to the incoming rays [see figure L2].

Retro-reflectors mounted on a spacecraft allow its distance from Earth (its range) to be accurately calculated, using a laser beam directed towards it. The **Apollo** programme left a number of retro-reflector packages on the Moon which have been used to determine its distance, and the variations in that distance, to accuracies of a few tens of centimetres.

When retro-reflectors take the form of three equilateral triangles joined to make an open pyramidal shape, they are called corner-cube reflectors. (Radar retro-reflectors of this design can be mounted on the masts of boats to assist in their detection).

LASER RING GYRO

See **laser gyro**.

LAUNCH

The procedure by which a **spacecraft** or **launch vehicle** is propelled into an **orbit** or **trajectory**; the moment of **lift-off** (also called 'blast-off').

Most launches are made from a **launch platform** on the surface of the Earth, but the term can be used in connection with any **planetary body**.

Spacecraft are also 'launched' from the **payload bay** of the American **Space Shuttle** once it has reached **low Earth orbit**.
[See also **flight**.]

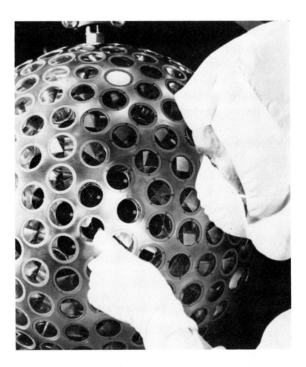

Figure L2 Lageos 1 satellite designed for experiments on accurate **range** determination using the **laser ranging retro-reflector**s which cover its surface. [Aeritalia]

LAUNCH CAMPAIGN

The period during which a **launch site** and a **launch vehicle** are prepared for a **launch**, typically including the delivery of the vehicle **stage**s to the launch site, their assembly, **payload integration** and all other activities leading to the **countdown** and **lift-off**.
[See also **mission**.]

LAUNCH COMPLEX

The general area surrounding a **launch pad**; a self-contained facility for launching **launch vehicle**s. Sometimes called a **launch site**, but there is usually more than one launch complex (and more than one launch pad) on a launch site.

LAUNCH CONTROL CENTRE (LCC)

A building from which the **launch** of a **launch vehicle** is controlled and monitored. When rocket launches were simpler affairs, the equivalent building would have been termed a 'firing room' or 'blockhouse', especially in a military context.

Most LCCs are known colloquially as 'mission control', since the **mission** of the typical launch vehicle is completed with the injection of its **payload** into a particular **trajectory**. In the case of a manned **NASA launch from Kennedy Space Center**, a distinction is made between LCC and MCC (**mission control centre**): once the launch vehicle has cleared the tower, responsibility for the mission is handed from the LCC at KSC to the MCC in Houston.

LAUNCH ESCAPE TOWER

See **escape tower**.

LAUNCH PAD

(i) A platform from which a **launch vehicle** is launched. Also called a **launch platform** [see figure L3].

Figure L3 Space Shuttle, carried by the **crawler** transporter, arrives at the **launch pad**. The opening of the **flame trench** and the **flame deflector** are visible beyond the Shuttle/Crawler combination. [NASA]

(ii) The platform *and* its immediate surroundings (e.g. **service structure**, **flame trench** and miscellaneous **ground support equipment**).

LAUNCH PLATFORM

(i) A structure from which a **launch vehicle** is launched; an integral part of a **launch pad** (sense (ii)).

(ii) An off-shore platform used for **rocket** launches (e.g. **San Marco platform**).

[See also **mobile launch platform**.]

LAUNCH PROFILE

(i) The shape of a **launch vehicle**'s **trajectory** with reference to the surface of the Earth; a graphical representation of the vehicle's **altitude** against its distance **downrange** from the **launch site** [see figure L4].

(ii) Any general description of a launch vehicle's **flight** parameters, including **range**, height, velocity, etc.

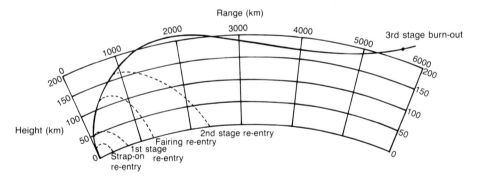

Figure L4 Launch profile for an **Ariane** 4 launch vehicle.

LAUNCH SHROUD

See **shroud**.

LAUNCH SITE

(i) A facility engaged in rocket launches; sometimes called a 'spaceport' [see **Baikonur Cosmodrome**, **Guiana Space Centre**, **Jiuquan**, **Kennedy**

Space Center (KSC), Northern Cosmodrome, San Marco platform, Tanegashima, Vandenberg Air Force Base (VAFB), Xichang.]

(ii) The general area surrounding a **launch pad**, otherwise known as a **launch complex**.

LAUNCH VEHICLE

Any vehicle designed to carry **payload**s into space.

A launch vehicle can be categorised in many different ways: it may be expendable or reusable (or a combination of the two); it may use **solid propellant** or **liquid propellant** or both; it may be capable of launching a payload to a **sub-orbital** or interplanetary **trajectory, low Earth orbit** or a **transfer orbit**; it may have one **stage** or several; it may or may not be **man-rated**.

[See also **expendable launch vehicle, heavy lift vehicle, aerospace vehicle, orbit, single stage to orbit, Ariane, Atlas, Blue Streak, Delta, Energia, Europa, Jupiter, H, Long March, N, Proton, Redstone, Saturn 1B, Saturn V, Scout, Space Shuttle, Thor, Titan.**]

LAUNCH WINDOW

A limited period of time during which the **launch** of a particular **spacecraft** may be undertaken. For launches into Earth **orbit** the main constraints are the desired orbit, the point of **injection** into that orbit, and the possibility of a collision with other orbiting spacecraft or **orbital debris**. For launches to interplanetary trajectories, the time of launch is constrained by the relative positions of the planets, at the time of both launch and **encounter** (i.e. the spacecraft must be aimed at a point in space that the planet will have reached by the time of encounter). A spacecraft launched outside the launch window would require more **propellant** and/or more time to reach its target.

LAUNCHER

A colloquial term for **launch vehicle**.

LAUNCHER RELEASE GEAR

See **hold-down arm**.

LEM

An acronym for lunar excursion module. See **Apollo**.

LENS ANTENNA

A type of antenna that focuses RF radiation in a manner analogous to

an optical lens: the longer lengths of **waveguide** or dielectric medium reduce the velocity of the propagating wave, which alters the form of the wavefront [see figure L5]. Variations are known as dielectric and **microwave** lenses.
[See also **antenna**.]

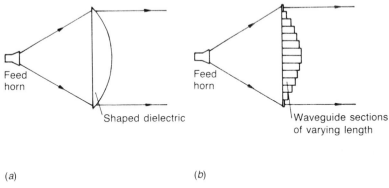

(a)   (b)

Figure L5 (a) A dielectric **lens antenna** and (b) a **waveguide** lens antenna.

LEO

See **low Earth orbit**.

LEWIS RESEARCH CENTER

See **NASA**.

LF (LOW FREQUENCY)

See **frequency bands**.

LHCP (LEFT-HAND CIRCULAR POLARISATION)

See **polarisation**.

LIBRATION PERIOD

A period of oscillation of a satellite about a mean position on the **geostationary orbit**, caused by solar radiation pressure [see **solar wind**].
[See also **equilibrium point**.]

LIBRATION POINT

See **Lagrange point**.

LIFE SUPPORT

The theory and practice of sustaining life in an environment or situation in which the human body is incapable of sustaining its own natural functions. In space this principally involves protection against the vacuum, harmful radiation and temperature extremes, and the provision of a breathable atmosphere at a comfortable atmospheric pressure, a thermal control system and a water supply.
[See also **portable life support system, spacesuit, decompression.**]

LIFE SUPPORT SYSTEM

Any collection of equipment designed to provide **life support** (e.g. that within the American Space Shuttle **orbiter** is called an **environmental control and life support system** (ECLSS)). A mechanical or biological system that recycles air, water and food is sometimes called a closed-ecology life support system (CELSS).
[See also **portable life support system, spacesuit.**]

LIFETIME

The length of time for which a device, vehicle or system is intended to perform its function. A distinction is made between its 'design lifetime' and its 'operational lifetime'. The design lifetime is the period over which it is designed to be *capable* of operation, whereas the operational lifetime is the period over which one may actually plan to use it (which may be less), or the period over which it remains in **operational** status (which may be more). For example, a **satellite** with a design lifetime of 10 years may be intentionally employed for only 8 years or it may be allowed to operate for 12 years.

The 'design lifetime' of different types of **spacecraft** is dependent upon different factors. That of satellites in **geostationary orbit** (e.g. **communications satellites, meteorological satellites**, etc) is governed mainly by the amount of **propellant** carried for **station keeping**: once this is depleted, the satellite can no longer hold its **orbital position** and begins to **drift**. Before this occurs, it is generally moved out of orbit to make way for another spacecraft [see **graveyard orbit**]. Communications satellites typically have design lifetimes between 7 and 10 years, although longer periods are being proposed. The lifetime of satellites in **low Earth orbit** is constrained chiefly by the propellant that can be carried for orbital adjustments and the **perigee** of the orbit, which determines the degree of **atmospheric drag**. The lifetime of **astronomical satellites** which carry infrared detectors is generally constrained by a limited supply of cryogenic coolant. Although the design lifetime of the constituent parts of an irretrievable satellite should at least equal the desired operational lifetime of the whole vehicle, components can be allowed to fail if a **redundancy** provision has been made.

A reusable **launch vehicle**, such as the American Space Shuttle **orbiter**, typically has an operational lifetime longer than that of some of its component parts (e.g. **Space Shuttle main engine**s are intended to be used for only about ten flights).
[See also **beginning of life (BOL), end of life (EOL)**.]

LIFT

The aerodynamic force produced by a **lifting surface**, perpendicular to the air-flow and opposing gravity.
[See also **lifting body, aerospace vehicle**.]

LIFT-OFF

The moment of **launch**; the initial vertical movement of a **launch vehicle** from the surface of a **launch platform**. More colloquially termed 'blast-off'.

LIFTING BODY

An aerodynamic vehicle, usually an unpowered glider, used—amongst other things—to evaluate proposed designs for future **aerospace vehicle**s.

LIFTING SURFACE

Any surface which provides **lift** (e.g. a wing).
[See also **aerospace vehicle**.]

LIGHT (VELOCITY of)

The velocity at which **electromagnetic radiation** travels in **free space** (i.e. 2.998×10^8 m s^{-1}).
[See also **light year**.]

LIGHT BAFFLE

See **baffle**.

LIGHT YEAR

The distance travelled by light in one mean solar year (i.e. 9.46×10^{15} m), where the velocity of light is 2.998×10^8 m s^{-1} and 1 year is 3.156×10^7 s; a unit of distance used in astronomy. Subdivisions: light minute and light second (e.g. 'the Sun is 8.3 light seconds from Earth').
[See also **parsec**.]

LINE HEATER

A heating device wound onto **propellant** lines to maintain the propellant above its freezing point. The **hydrazine** liquid propellant commonly used in **spacecraft** reaction control systems is particularly prone to low temperatures in the extremities of the system where the volumes are low: in the lines, valves and **thrusters**, the latter of which are on the spacecraft exterior.
[See also **heater, thermal insulation, thermal control subsystem.**]

LINEAR POLARISATION

See **polarisation**.

LINER

(i) A material placed between the **motor case** and the **propellant** in a **solid rocket motor** [figure R5]. See **propellant liner, flame inhibitor**.

(ii) A thin layer of adhesive used to bond **solid propellant** to a motor case or to a layer of **thermal insulation** or thermal blanket.

(iii) An ablative material lining the inner wall of a **combustion chamber** or **nozzle** to reduce heat transfer to the wall or nozzle.

[See also **rocket motor, ablation.**]

LINK BUDGET

A calculation of signal power in a communications link design which indicates whether or not a signal will be successfully received. A link budget is the 'profit and loss account' of the transmitted signal. 'Profit' is accrued by the amplification of the signal, either in the satellite **transponder** or in the **earth station**'s **amplifier chain**, and from the inherent **gain** of the **spacecraft** and earth station **antennas**. 'Losses' arise from the signal's distance of travel: on its path through space, through Earth's atmosphere and in the communications hardware. The link budget constitutes a 'balance sheet' which combines these profits and losses; the resultant **margin** in signal power is the balance.
[See also **free space loss, atmospheric attenuation, I^2R loss.**]

LINK MARGIN

The amount by which the **RF power** in a communications **link budget** is above that required to achieve the performance specified for the link under **nominal** conditions. The point where there is no **margin** will depend on the quality standards set; if the power drops below this point, there is said to be a 'negative margin'. Usually expressed in **dB**.

LIQUID APOGEE ENGINE (LAE)

A **rocket engine** used to transfer a satellite from **geostationary transfer orbit** (GTO) to **geostationary orbit** (GEO), fired when the spacecraft reaches the point in the GTO ellipse furthest from the Earth, the **apogee**. It is usually part of a **combined propulsion system** (combining apogee boost and reaction control functions), and uses **liquid propellant**s.
[See also **apogee kick motor (AKM)**, **perigee kick motor (PKM)**, **rocket motor**, **reaction control system (RCS)**.]

LIQUID COOLING AND VENTILATION GARMENT (LCVG)

See **spacesuit**.

LIQUID HYDROGEN

A widely used **cryogenic propellant**; a liquid **fuel** with a boiling point at -252.8 °C and a melting point at -259.2 °C, widely used with **liquid oxygen** as the **oxidiser**. Commonly abbreviated to 'LH$_2$'.
[See also **slush hydrogen**, **liquid propellant**, **rocket engine**.]

LIQUID OXYGEN

A widely used **cryogenic propellant**; a liquid **oxidiser** with a boiling point at -182.98 °C and a melting point at -218.76 °C, widely used with **kerosene** or **liquid hydrogen** as the **fuel**. Commonly abbreviated to 'LO$_2$' or 'LOX'.
[See also **liquid propellant**, **rocket engine**.]

LIQUID PROPELLANT

The 'working fluid' expelled from a rocket engine to provide **thrust**. There are two main types of liquid propellant: 'storable' and 'non-storable', the former being a storable liquid at terrestrial environmental temperatures and the latter needing various degrees of cooling below these temperatures to render it liquid. The non-storable propellants are otherwise termed 'cryogenic' after the branch of physics concerned with very low temperatures [see **cryogenic propellant**]. Liquid propellants are generally more difficult to store and handle than the solids, and even some of the storable liquids are, to some extent, hazardous, being highly reactive and toxic. The liquid engine is, however, more flexible in its operation [see **rocket engine**] and liquid propellants typically exhibit a higher **specific impulse** than solids.
    A liquid propellant is either a **fuel** or an **oxidiser**. Examples of

storable liquid oxidisers widely used for rocket propulsion in the past are red fuming nitric acid (RFNA) and hydrogen peroxide (H_2O_2). Currently more common is nitrogen tetroxide (N_2O_4).

As for storable liquid fuels, the methyl and ethyl alcohols, used in early rockets such as the **V-2** with liquid oxygen as the oxidiser, have since been replaced by hydrocarbons such as kerosene, which provide a more efficient mixture (kerosene was used with liquid oxygen in the first stage of the **Saturn V**). Two fuels which have been used with nitric acid are hydrazine (N_2H_4) and unsymmetrical dimethylhydrazine (UDMH: $(CH_3)_2N.NH_2$). Hydrazine can also be used with hydrogen peroxide or N_2O_4, both of which are **hypergolic** combinations; UDMH is widely used with N_2O_4 (e.g. in the first two stages of the **Ariane** launch vehicle).

Many **reaction control systems** use hydrazine as a **monopropellant** (i.e. without an oxidiser), but the hypergolic combination of nitrogen tetroxide and monomethyl hydrazine (MMH : $CH_3NH.NH_2$), used in the liquid **bipropellant** system, is becoming equally common [see **combined propulsion system**].

[See also **propellant, solid propellant, liquid hydrogen, liquid oxygen**.]

LITHIUM HYDROXIDE CANISTER

See **environmental control and life support system (ECLSS)**.

LIVE STAGE

A launch vehicle **stage** containing **propellant** and capable of powered flight (cf a 'live round' of ammunition).
[See also **dummy stage**.]

LNA

See **low-noise amplifier**.

LNB

An acronym for low-noise block-downconverter. See **low-noise converter**.

LNC

See **low-noise converter**.

LOAD

(i) Mechanical: another name for a force exerted on a component, mechanism or structure—'structural load'. Mechanical loads can be either static or dynamic: static loads are exerted on a stationary structure; dynamic loads on a structure in motion.

(ii) Electrical: a device which dissipates the power of a **signal**, often used to terminate a **transmission line** to absorb unwanted signal power on that line. 'A load' is similar to a very-high-loss **attenuator**.

(iii) Electrical: the **energy** delivered by, or that required by, an item of power subsystem equipment—colloquially termed the 'power load'.

[See also **load path, thrust structure, payload, power supply, power bus**.]

LOAD PATH

The path through which mechanical loads are transmitted in a structure.

For example, knowledge of the load path helps to ensure that a spacecraft structure is capable of surviving the mechanical loads imparted upon it by a **launch vehicle**. The load path between a launcher and a satellite is via the launch vehicle adapter ring to a rigid **thrust structure** which, along with a number of platforms and panels, constitute the 'primary structure' of the satellite. The remaining structural elements, such as support brackets for propellant tanks, mountings for **antenna** reflectors, **feed** assemblies, etc, fit into the category of 'secondary structure', thus completing the load path from subsystem equipment, through secondary and primary structures, to the adapter ring.

LOCAL OSCILLATOR (LO)

A device which provides a stable, fixed-frequency source of **radio-frequency** energy used in telecommunications equipment (e.g. in a spacecraft **transponder**: see **mixer, downconverter, upconverter**). Alternatively known as a 'carrier generator'.

When used to supply an RF **beacon**, or a similar device whose output is used for transmission in its own right, rather than for frequency conversion, this type of device tends to be called simply a 'frequency source'.

LONG MARCH

A Chinese **launch vehicle** developed, in three variants, during the late 1960s and early 1970s. The first two **liquid propellant** stages of the Long March 1 (otherwise known as the CZ-1) were developed from the CSS-2 missile, to which was added a third **solid propellant** stage. It was introduced in 1970. The CZ-2 (originally called the FB-1) was a two-stage vehicle first flown in July 1975 and used as a satellite launcher until 1983. The CZ-3, which made its first orbital flight in January 1984, is composed of the two liquid stages of the FB-1 using **nitrogen tetroxide/UDMH** and a third stage using **liquid oxygen/liquid hydrogen**. It was offered to the West as the commercial Long March 3 launch vehicle in the late 1980s. From its launch site near **Xichang** (at 27.9°N, southwest of Chengdu) it has a **payload** capability of 1400 kg to **geostationary transfer orbit** (GTO).

LONGERON

A main longitudinal structural brace in a **launch vehicle** or **spacecraft** structure.
[See also **stringer, airframe**.]

LOS

An acronym for loss of signal. Used typically when a **communications** link with a **spacecraft** has been interrupted (e.g. due to blockage by a **planetary body**, radio **interference** or during **re-entry** [see **S-band blackout**].). The opposite of 'acquisition of signal' (AOS).
[See also **signal, carrier**.]

LOUVRE(S)

A device used in the thermal control of a spacecraft to shield or expose a **radiator** surface [see figure L6]. The angle of the louvres, which resemble a domestic venetian blind, is varied by a bimetallic spring (or, less commonly, by the expansion and contraction of a fluid).

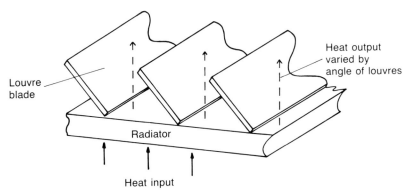

Figure L6 Thermal **louvres** used in the thermal control of a spacecraft.

Thermal louvres are not generally used on **communications satellite**s in geostationary orbit, since other methods of thermal control are more effective. They are, however, sometimes used for spacecraft in **low Earth orbit**, and those on interplanetary trajectories.

LOW EARTH ORBIT

A nominally circular **orbit** of low altitude (typically less than 1000 km), and short period (approximately 90 minutes). Commonly abbreviated to LEO.

This type of orbit, which includes **polar orbit**, is predominantly occupied by **remote sensing**, military and other non-communications satellites which require (i) a close proximity to the Earth's surface for **imaging** and (ii) a short orbital period (a couple of hours or so) to allow rapid 'revisits'. Since satellites in low Earth orbits move quickly across the sky and soon disappear below the observer's horizon, they are not very useful for telecommunications applications (**communications satellite**s tend to occupy the much higher **geostationary orbit**).

Most manned spaceflight activities have been confined to LEO and it is here that future manned **space station**s will reside. The American **Space Shuttle** is designed to deliver payloads to LEO.

[See also **orbital decay**, **drag compensation**, **re-visit capability**.]

LOW GRAVITY

A term used to describe an environment where the acceleration due to gravity has a value less than that on Earth (e.g. the 1/6g of the Moon). [See also **gravity**, **microgravity**, **gravitational anomaly**, **perturbations**.]

LOW-NOISE AMPLIFIER (LNA)

A type of amplifier designed to deliver a good **signal-to-noise ratio**. In a spacecraft **communications payload** the LNA is usually one of the 'pre-amplifiers' preceding the **channel filter**. As part of a receiving **earth terminal** the amplifier may be combined with a frequency converter: it is then termed a **low-noise converter** (LNC).

LOW-NOISE CONVERTER (LNC)

A **low-noise amplifier** combined with a frequency converter, usually mounted at the focus of an **earth terminal** antenna [see figure L7]. The unit amplifies the high-**frequency** satellite signal and converts it to a lower **intermediate frequency** (IF). Sometimes called an LNB, or low-noise block-downconverter, although LNBs tend to be very **wideband** devices (i.e. they convert the whole of an available band), whereas LNCs

can be quite narrow band.
[See also **frequency bands**.]

Figure L7 A **low-noise converter** (LNC) mounted at the focus of an **earth terminal** antenna. [Mark Williamson]

LOW-PASS FILTER

A **filter** with a high-**frequency** cut-off which allows only low frequency signals to pass.
[See also **band-pass filter**.]

LOW-POWER DBS

See **direct broadcasting by satellite**.

LOX

A commonly used abbreviation for **liquid oxygen**.

LUNA

The name given to a series of Soviet spacecraft designed for lunar research: Luna 9 made the first 'soft landing' on another **planetary body** on 3 February 1966.

LUNAR 'LOPE'

A form of locomotion, developed by the **Apollo** lunar **astronaut**s, which proved particularly effective in the 1/6g environment of the Moon; an undulating 'one-legged skip'.

LUNAR MODULE

See **Apollo**.

LUNAR ROVING VEHICLE (LRV)

One of three electrically powered motor vehicles used to extend the area of the Moon which could be explored during the last three **Apollo** lunar missions (Apollos 15, 16 and 17) [see figure A7]. Variously called a 'lunar rover', 'Moon car' or 'Moon buggy', the LRV was stored in the **descent stage** of the lunar module in a package of volume 0.85 m^3 and unfolded to a length of 3.1 m and a width of 1.82 m. It was capable of carrying two **astronaut**s, their equipment and lunar samples at a maximum speed of 14 km h^{-1} (8.7 mph).

LUNI–SOLAR GRAVITY

The collective gravitational attraction of the Moon and Sun; one of the mechanisms that results in the perturbation of a spacecraft's **orbit**.

The varying effect of the gravitational fields of the Sun and Moon, as their positions relative to the Earth change, causes a tilting of a satellite's orbit plane and a north–south oscillation of the spacecraft's position as seen from the Earth. This tendency to wander is corrected by the spacecraft's **orbital control** subsystem.

[See also **perturbations, equilibrium point, triaxiality, solar wind**.]

M

MAC

See **multiplexed analogue component**.

MACH NUMBER

The ratio of the speed of a body in a particular medium to the speed of sound in that medium, such that 'mach 1' corresponds to the speed of sound (named after the Austrian physicist Ernst Mach (1838–1916)). Applicable both to the motion of a body in a fluid and the motion of a fluid in a body (e.g. a rocket **exhaust** in a **rocket engine**). If the mach number is less than 1 the flow is 'subsonic'; if it is greater than 1 the flow is 'supersonic'; if it is 5 or greater it can also be called 'hypersonic'. [See also **nozzle**.]

MAGNETOMETER

A device for measuring the intensity and orientation of a magnetic field. Usually carried by planetary **probe**s to measure planetary magnetic fields. See **magnetosphere**.

MAGNETOPAUSE

See **magnetosphere**.

MAGNETOSPHERE

A region surrounding the Earth in which charged particles from the Sun are shepherded by the Earth's magnetic field and manipulated by the **solar wind**. The magnetosphere takes its general form from the lines of force of the Earth's magnetic field, which 'loop' from pole to pole like those from an ordinary bar magnet. The inner region is a 'geomagnetic trap' for charged particles, which spiral round the lines of force from pole to pole, while the outer region is swept into a 'geomagnetic tail' by the solar wind [see figure V1]. The edge of the magnetosphere, known as the magnetopause, is compressed to an altitude of about 70 000 km on the sunward side and 'blown' beyond the Moon's orbit on the opposite side.

Any planet with a magnetic field should exhibit a similar magnetosphere—in fact Jupiter's has been shown to be particularly impressive.

[See also **Van Allen belts, ionosphere, magnetometer.**]

MAGNETO-TORQUER

A device which makes use of a planet's magnetic field to stabilise a satellite. It is based on the principle that a freely suspended current-carrying coil aligns itself with the local magnetic field. Small disturbances in the satellite's attitude will be damped out as a battery-fed coil, mounted in the structure of the satellite, acts to realign itself with the planet's field. Since the effect varies with the strength of the field, the method is best suited to applications in low, rather than high, Earth orbits.

[See also **stabilisation.**]

MAJOR AXIS

The long dimension of an ellipse described by a line passing through both foci. The major axis of an elliptical **orbit** about a **planetary body** is the line passing through the body with the **periapsis** at one end and the **apoapsis** at the other. The major axis of an **elliptical antenna** is the longest dimension; the minor axis is the shortest dimension.

MAN-RATED

An attribute of a **launch vehicle** classified as suitable (i.e. sufficiently safe) to carry a **manned spacecraft**.

MAN-TENDED SPACECRAFT

A **spacecraft** which usually operates unmanned but is designed to be visited by **astronaut**s for replenishment of **consumable**s; a spacecraft designed for human habitation, but only on a short period basis (cf a **manned spacecraft**). Otherwise known as a 'man-tended **platform**'.

[See also **unmanned spacecraft.**]

MANEUVER

American spelling of **manoeuvre**.

MANIPULATOR ARM

See **remote manipulator system**.

MANNED MANOEUVRING UNIT (MMU)

A self-contained propulsive device carried in the **payload bay** of the American **Space Shuttle** and used by **astronaut**s for untethered **extra-vehicular activity** (EVA) [see figure M1]. It is, in effect, a manned space vehicle with its own power supply, **propulsion system**, **inertial guidance** system, controls and displays. Its **propellant** is gaseous nitrogen and it uses 24 fixed **thrusters** to provide position control with six **degrees of freedom**. When carried, MMUs are stowed in 'flight support stations' aft of the crew compartment. A forerunner to the MMU, called the astronaut manoeuvring unit (AMU), was tested aboard the **Skylab** space station in the mid 1970s.

Figure M1 Astronaut Bruce McCandless flying a **manned manoeuvring unit** (MMU) above the **Space Shuttle** payload bay on mission STS 41-B [see also figure S16]. Mounted in front of the MMU is a docking apparatus which can be used to manoeuvre damaged satellites into the payload bay for repair. Visible in the **payload bay** is the **sunshield** protecting the Westar VI communications satellite. [NASA]

MANNED SPACECRAFT

Any **spacecraft** designed to carry a **crew**. Any other spacecraft is either an **unmanned spacecraft** or a **man-tended spacecraft**.
[See also **Apollo, Gemini, Mercury, Mir, Salyut, Skylab, Soyuz, Spacelab, Space Shuttle, Space Station, Vostok, Voskhod**.]

MANOEUVRE

Any change made to the **attitude, orbit** or **trajectory** of a **spacecraft, launch vehicle** or **aerospace vehicle**.
[See also **plane change, mid-course correction, attitude control, orbital control, orbital injection, de-orbit, orbital relocation**.]

MARGIN

An additional amount of a resource, over and above the minimum required, included as a contingency reserve (e.g. a '**propellant** margin', a '**power** margin', etc) [see **propellant budget, power budget**]. Since all spacecraft **hardware** and **consumables** represent mass which must be launched, the magnitudes of the margins have to be very carefully controlled [see **mass budget**]. An amount which is less than the minimum required is referred to as a 'negative margin'.
[See also **link margin, beginning of life (BOL), end of life (EOL)**.]

MARINER

The name given to a number of American planetary exploration **probes** launched towards Mars, Venus and Mercury.

MARSHALL SPACE FLIGHT CENTER

See **NASA**.

MASCON

An abbreviation for 'mass concentration'. Usually refers to a concentration of mass detected through a change in a spacecraft's **orbital parameters** as it circles a **planetary body**. Otherwise referred to as a 'gravitational anomaly'.
[See also **perturbations**.]

MASER

An acronym for **microwave** amplification by stimulated emission of radiation. See **laser**.

MASS BUDGET

A method of accounting for the mass of a spacecraft or subsystem with respect to the launch-mass capability of the intended **launch vehicle**.

During a spacecraft project, each subsystem engineering discipline aims to design its hardware within its allotted portion of the mass budget. The sum of subsystem masses allows the overall spacecraft mass to be monitored throughout the project. This sum would usually incorporate a '**balance mass**', typically about 1% of the total, which must be allowed for when balancing the total spacecraft in preparation for launch. As it stands, this budget represents the 'dry mass' of the spacecraft: the addition of propellant for orbital injection and station keeping can more than double the mass.

[See also **power budget, propellant budget, link budget**.]

MASS-LIMITED

Jargon: 'limited in terms of the allowable mass'. The mass of anything launched into space is limited by the **performance**, or **payload** capability, of the **launch vehicle**. The mass of all components and **subsystem**s is therefore a major constraint in any **spacecraft** design [see **mass budget**.]. By way of illustration, a **communications satellite** capable of providing sufficient **power** to operate more **transponders** than the mass budget would allow it to carry would be 'mass-limited'.

[See also **power-limited, materials**.]

MASS RATIO

(i) The ratio of the mass of a **launch vehicle**'s **payload** to the mass of the total vehicle at **lift-off**. Also called 'payload mass ratio' and 'payload fraction'. Typical figures for a three-stage **expendable launch vehicle** are about 1% to **geostationary orbit** and 2–3% to **low Earth orbit**.

(ii) The ratio of a launch vehicle's initial mass at lift-off to the mass of the vehicle without **propellant** (i.e. its **dry mass**). Also called 'propellant mass ratio' and 'propellant mass fraction'.

MATE/DE-MATE DEVICE

A framework structure used to mount a Space Shuttle **orbiter** on the **Shuttle carrier aircraft** or remove it from it.

MATERIALS

One of the most important characteristics for materials used in spacecraft is low density for high structural strength (a high strength-to-weight ratio), due to the need to limit the launch mass of the spacecraft.

However, materials are chosen to match the particular application and other factors are also important (e.g. high stiffness-to-weight ratio, low thermal expansion coefficient, etc). The following materials are used to varying degrees: **aluminium, beryllium, titanium, carbon composite, ceramics, ceramic–matrix composite, metal–matrix composite**.

MAX-Q

An abbreviation for maximum dynamic pressure—see **dynamic pressure**.

MAXIMUM DYNAMIC PRESSURE

See **dynamic pressure**.

MCC

An acronym for (i) **mid-course correction** and (ii) **mission control centre**.

MECO

An acronym for Main Engine Cut-Off, specifically the point at which the **Space Shuttle main engine**s cease **combustion** during a **launch** of the American **Space Shuttle** (at about $T+8$ minutes, 40 seconds).

MEDIUM-POWER DBS

See **direct broadcasting by satellite**.

MERCURY

A series of six American **manned spacecraft** launched in the early 1960s, largely to show that an **astronaut** could survive a spaceflight, control a spacecraft, etc. The one-man **capsule** had no room for an ejection seat, but was topped by a **launch escape tower**. **Attitude control** in orbit was accomplished using **hydrogen peroxide** thrusters. A **retro-pack** comprising three **solid propellant** rocket motors was secured to the **heat shield** by metal straps and jettisoned prior to **re-entry**. The shield itself was detached before **splashdown** to expose a rubberised glass-fibre pneumatic cushion, known as a 'landing bag', to absorb the impact.

The first two manned Mercury capsules were launched by a **Redstone** launcher onto **sub-orbital** trajectories: the first American in space was Alan Shepard, launched 5 May 1961 on the world's second manned spaceflight (of just over 15 minutes duration) [see **Vostok**]. The remaining four missions were launched into a **low Earth orbit** by **Atlas**

boosters: the first American *in orbit* was John Glenn, on the third Mercury flight, launched 20 February 1962 (duration 4 h 55 min). Only six of the seven pilots chosen for the Mercury flights (known as 'The Mercury Seven') actually flew in a Mercury capsule: Deke Slayton was 'grounded' due to a suspected heart complaint until 1975 when he flew on the **Apollo–Soyuz test project**.
[See also **Gemini, Apollo**.]

MET

See **mission elapsed time**.

METAL–MATRIX COMPOSITE (MMC)

A material composed of a metal matrix reinforced by threads or fibres of a metallic or ceramic material. Example matrices are aluminium, lead, copper, magnesium and titanium; possible fibres are graphite, boron, silicon carbide, tungsten, molybdenum and alumina. MMCs have high strength and stiffness at high temperatures, good dimensional stability and high thermal and electrical conductivities. They have lower thermal expansion coefficients than the thermosetting polymers and exhibit good resistance to radiation damage, little outgassing and no low-temperature brittleness.
[See also **composites, carbon composite, ceramic–matrix composite, pre-preg, materials**.]

METEOROID

See **micrometeoroid**.

METEOROLOGICAL SATELLITE

A specialised class of **remote-sensing** satellite concerned primarily with imaging weather patterns, usually in the visible and infrared parts of the spectrum. The majority of 'metsats' occupy positions in **geostationary orbit**, from where a **constellation** of three satellites can provide full Earth coverage, except for the polar regions.

METSAT

A colloquial abbreviation for **meteorological satellite**.

MF (MEDIUM FREQUENCY)

See **frequency bands**.

MGSE

An acronym for mechanical ground support equipment. See **ground support equipment**.

MHz

Abbreviation for megahertz—see **hertz, frequency bands**.

MICROGRAVITY

A term used to describe an environment in which the acceleration due to gravity is approximately zero; otherwise termed **weightlessness**. The only accelerations are those produced, for example, by **station-keeping** manoeuvres, the movements of mechanical devices (e.g the deployment of equipment on **boom**s), and the movements of any **crew**-members.

The microgravity environment of **low Earth orbit**, particularly, is useful for certain types of manufacture (e.g. that of semiconductor crystals, which tend to grow larger and with fewer defects than they would on Earth).

[See also **gravity, low gravity, gravitational anomaly, perturbations**.]

MICROMETEOROID

Any small **celestial body** of natural origin, about the size of a grain of sand, or smaller; a 'small meteoroid'—the distinction is open to discussion.

Once the body has collided with the Earth, another **planetary body** or a spacecraft, it becomes a meteorite or micrometeorite (a meteor is a meteoroid made visible by frictional heating in the atmosphere). Depending on their size and relative velocity, micrometeoroids can degrade spacecraft surfaces (**thermal insulation, solar array**s, etc) and, at worst, puncture pressurised **cabin**s or **module**s. However, depending on the **orbit**, collision with man-made objects can be more of a problem—see **orbital debris**.

MICROWAVE

The section of the **electromagnetic spectrum** between 0.001 and 0.3 m ($1 \times 10^9 - 3 \times 10^{11}$ Hz; 1–300 GHz). See **radio frequency (RF)**.

MICROWAVE LENS

See **lens antenna**.

MICROWAVE LINK

A terrestrial communications system utilising **parabolic antenna**s, operating at microwave frequencies $(1 \times 10^9 - 3 \times 10^{11}$ Hz), typically mounted on high towers to increase their range.

MID-COURSE CORRECTION (MCC)

An adjustment made to a spacecraft's **trajectory** at an intermediate point in its flight, particularly those travelling between planetary bodies. It is practically impossible to inject a spacecraft into a desired orbit around another **planetary body** using only one motor '**burn**'. MCCs allow the initial trajectory to be improved upon as the spacecraft's actual trajectory is determined throughout the flight.
[See also **parasitic station acquisition**.]

MID-DECK

The part of the American Space Shuttle **orbiter** directly below the **flight deck** which houses the living quarters, storage lockers and, on some missions, an **airlock** which opens into the **payload bay** .

MINOR AXIS

The shortest aperture dimension of an **elliptical antenna**.
[See also **major axis**.]

MIR

A Soviet **space station** launched on 20 February 1986 by a **Proton** launch vehicle. The station, the name of which is translated as 'peace', has a total habitable volume of about 100 m³ and is similar to the **Salyut** 7 space station. It has a number of **docking port**s to accept visiting **Soyuz** 'ferries' and Progress re-supply 'tankers'. An astrophysics **module** called 'Kvant' ('Quantum') was added in April 1987 and further expansion is planned.
[See also **Skylab**.]

MISSION

A specific task assigned to a **spacecraft, launch vehicle** or **crew**. The duration of the mission of an **unmanned spacecraft** is measured in terms of its 'operational **lifetime**'; that of a **manned spacecraft** typically extends from **lift-off** to **touchdown** [see **mission elapsed time**]. The mission of a typical launch vehicle is completed with the injection of its **payload** into a designated **orbit** or **trajectory**.
[See also **mission specialist**.]

MISSION CONTROL

A collection of personnel and computer systems with the task of overseeing and controlling a **launch** or a space **mission**, whether manned or unmanned; the function of the 'mission controllers'.

MISSION CONTROL CENTRE (MCC)

A building from which a space **mission** is controlled and monitored [but see **launch control centre (LCC)**].

MISSION ELAPSED TIME (MET)

A measure of time since the beginning of a **mission** used by **NASA** in the control and monitoring of **manned spacecraft**. A mission clock is started at the moment of **lift-off** and stopped, in the case of the American **Space Shuttle**, at the instant the **orbiter** comes to rest after landing (on earlier missions at the moment of **splashdown**).

MISSION SPECIALIST

A category of person who is part of the crew of the American **Space Shuttle**, but has not been trained as an **astronaut**; typically an engineer or scientist responsible for an aspect peculiar to a particular **mission** (e.g. a scientific **payload** carried by the Shuttle **orbiter**).
[See also **payload specialist**.]

MIXED OXIDES OF NITROGEN

See **MON**.

MIXER

An electrical device in which two or more **input** signals are combined and filtered to produce a single **output** signal. The mixing or 'heterodyning' process is used, amongst other things, to convert one **radio frequency** to another in a satellite **communications payload**. See **upconverter, downconverter**.
[See also **local oscillator, filter**.]

MIXING PRODUCTS

In **telecommunications**, products of the mixing process in an **upconverter** or **downconverter** which may cause **interference** and are therefore undesirable. See **mixer**.

MLI

An acronym for multi-layer insulation. See **thermal insulation**.

mm-WAVE

Millimetre wave: 40 − 100 GHz. See **frequency bands**.

MMC

See **metal—matrix composite**.

MMH

An acronym for MonoMethyl Hydrazine ($CH_3NH.NH_2$), a storable liquid **fuel** used in **rocket engines**. See **liquid propellant**.

MMU

See **manned manoeuvring unit**.

MOBILE LAUNCH PLATFORM

In general, any **launch platform** which can be moved to a **launch pad** with a **launch vehicle** mounted upon it [see figure G5]. In particular, the Mobile Launcher Platform (MLP) upon which the American **Space Shuttle** is mounted for its transfer from the **vehicle assembly building** to the launch pad [figure C5]. The MLP is 41 m wide, 49 m long, 7.6 m high and weighs 3700 tonnes. It has three openings for the **Space Shuttle main engine** and **solid rocket booster** exhausts, two **tail service mast**s for **propellant** loading and electrical power, and a water-based **sound suppression system**. The MLP was originally used to transport **Saturn V** launch vehicles during the **Apollo** programme, when it also carried a launch tower or '**service structure**'. This is now a permanent fixture at the pad [but see **rotating service structure**].

The main advantage of a mobile platform is that a launch vehicle can be assembled under controlled environmental conditions with access to all ground-servicing equipment. It does, in addition, leave the launch pad free for actual launches, rather than vehicle **integration**.
[See also **crawler, crawlerway**.]

MOBILE-SATELLITE SERVICE (MSS)

A **satellite communications** system providing links between mobile **earth station**s and one or more orbiting spacecraft. MSS includes maritime (MMSS), aeronautical (AMSS) and land mobiles (LMSS).

MOBILE SERVICE STRUCTURE

See **service structure**.

MODEM

A contraction of 'MODulator–DEModulator'; the name usually given to a device for converting digital **data** to audio tones (and vice versa), typically used when connecting two computers via a telephone line.

MODULATION

In **telecommunications**, the process whereby a signal is superimposed upon a higher frequency **carrier** wave which 'carries' the signal until it is received and demodulated. Commonly referred to as 'modulating the carrier'. The modulating signal can be analogue or digital, originating from a telephone, TV camera, computer, etc, and is often referred to as the '**baseband**' signal.

There are two primary modulating techniques in general use: **amplitude modulation (AM)** and angle modulation, the latter of which includes **frequency modulation (FM)** and **phase modulation (PM)**. Prior to modulation by one of these techniques, a carrier may undergo a modulation of a different kind, a form of coding, e.g. **pulse code modulation** (PCM). Other methods less commonly used are differential pulse code modulation (DPCM), **Delta modulation** (DM) and adaptive versions of PCM, DPCM and DM (referred to as APCM, ADPCM and ADM, respectively).

An RF signal may also be 'modulated' onto an **electron beam**, for instance in a **travelling wave tube**.

MODULATOR

A device whose function is to modulate a signal onto a **carrier** wave. See **modulation**.

MODULE

(i) A self-contained unit or item which can be assembled and tested independently from other items and 'integrated with the rest of the **spacecraft, launch vehicle**, etc, at a later stage [see **integration**] [see figures M2 and M3]. The term is equally applicable to the smallest 'electronics modules' and the largest **subsystem** components (e.g. **payload module, service module, antenna module, docking module**).

(ii) A self-contained, and usually separable, part of a spacecraft or launch vehicle with a specified task (e.g. **command module, lunar module, crew** module, **habitable module, re-entry** module, **extension module**).

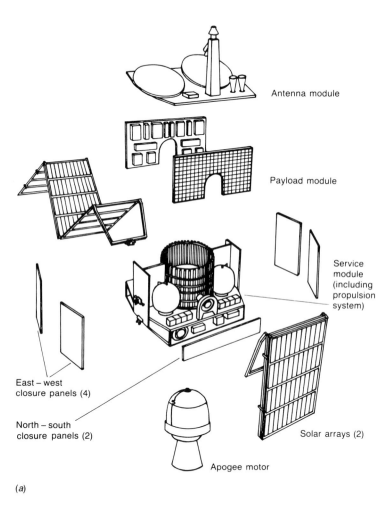

Antenna module

Payload module

Service
module
(including
propulsion
system)

East – west
closure panels (4)

North – south
closure panels (2)

Solar arrays (2)

Apogee motor

(a)

Figure M2 Modular design of (a) a **three-axis-stabilised** satellite and (b) a **spin-stabilised** satellite [see **module**].

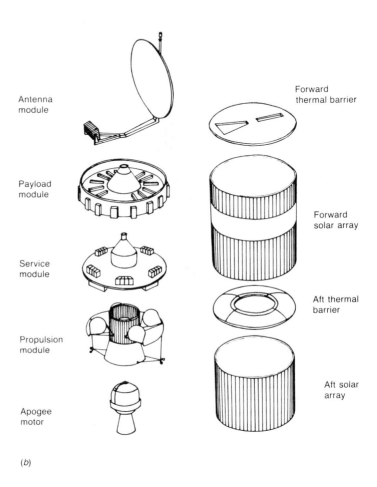

Antenna module

Payload module

Service module

Propulsion module

Apogee motor

Forward thermal barrier

Forward solar array

Aft thermal barrier

Aft solar array

(*b*)

Figure M2 (continued)

Figure M3 Integration of the ECS-1 payload and service **modules**. [ESA]

MOLNIYA

The name given to a series of Soviet **communications satellites**.

MOLNIYA ORBIT

An **elliptical orbit** which gives coverage of high latitudes due to its inclination of about 63° [figure M4]. Its **apogee** and **perigee** are approximately 39 000 km and 1500 km respectively. Since its period is 12 hours, it has two apogees giving two **coverage area**s on opposite sides of the Earth. A disadvantage is that of **atmospheric drag**, due to the low perigee, which also means that a spacecraft in this orbit will pass through the **Van Allen belts**.

The **orbit** was named after the Soviet 'Molniya' **communications satellite**s which pioneered its use. Full-time coverage is typically provided by three satellites synchronised in similar orbital paths so that each one contributes about eight hours of continuous service as it approaches and recedes from its apogee.

[See also **tundra orbit, geostationary orbit**.]

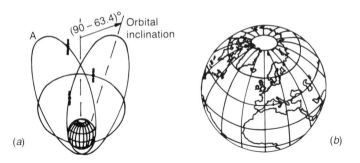

Figure M4 Satellites in a **Molniya orbit** [see also figure G4—ground track]. (*a*) Three-satellite Molniya system (as seen by a fixed observer); (*b*) view from apogee (A).

MOMENTUM BIAS

A property given to a spacecraft by an **attitude control** subsystem incorporating a **momentum wheel**; a system using a momentum wheel to stabilise a **three-axis-stabilised** spacecraft.

MOMENTUM DUMPING

The practice of reducing the angular momentum of a **momentum wheel** or **reaction wheel** which has reached its maximum allowable spin speed. Since any change in the wheel's spin speed causes the spacecraft to rotate, a **thruster** is fired as the wheel is decelerated to maintain the spacecraft's **attitude**. The process is alternatively known as 'off-loading momentum'.

MOMENTUM OFF-LOADING

See **momentum dumping**.

MOMENTUM WHEEL

A wheel mounted in a **three-axis-stabilised** spacecraft used to provide gyroscopic stability [figure M5].

A spinning momentum wheel, aligned with the spacecraft's north–south (**pitch**) axis, acts like a **gyroscope** to provide a resistance to perturbing forces in **roll** and **yaw** while allowing the body to rotate once every 24 hours about the pitch axis [see **spacecraft axes**]. An alternative system uses two wheels set at a slight angle to each other but symmetrically on either side of the N–S axis. Run together the wheels will act as a double-sized single wheel, but changing the speed of one will alter the balance to provide a degree of **attitude control**. These arrangements are examples of the so-called 'momentum bias' system.

Figure M5 A spinning **momentum wheel** (top cover removed). [Teldix]

A momentum wheel spins constantly in one direction with typical speeds between 6000 and 12 000 rpm, depending on the type of wheel, with a variation of plus or minus 10% of the nominal value for control adjustments. Compare this with the **reaction wheel**; both types of wheel are sometimes called inertia wheels.
[See also **momentum dumping**.]

MON

An acronym for 'mixed oxides of nitrogen', a **liquid propellant** comprising **nitrogen tetroxide** (N_2O_4) with an added proportion of nitric oxide (NO). The percentage (by weight) of added NO is indicated by a figure in a particular propellant designation, e.g. 'MON3'.

MONOCOQUE

A support structure which alone carries all or most of the structural stresses placed upon it (e.g. the **thrust structure** of a simple **spin-stabilised** satellite).

MONOMETHYL HYDRAZINE

See **MMH**.

MONOPROPELLANT

A chemical **propellant** which comprises a single component.

Liquid propellants can be either monopropellants or **bipropellant**s. **Solid propellant**s are considered by some as bipropellants because they have two components, but can also be regarded as monopropellants since they are handled and used as a single-component propellant: the **fuel** and **oxidiser** are mixed during the manufacturing process (cf liquid bipropellants which are stored separately and mixed in a **combustion chamber**).

Hydrazine is an example of a single-component monopropellant (stored as a liquid and decomposed to form a gas) which is widely used in **reaction control system**s.
[See also **hybrid rocket, hydrazine thruster**.]

MONOPULSE RF SENSOR

See **RF sensor**.

MOTOR

See **rocket motor**.

MOTOR CASE

The outer structure or pressure shell of a solid rocket motor (SRM) [figure R5]. The case is, in effect, a receptacle for the **propellant**, which is poured in during manufacture and 'moulds itself' to the contours of the case.
[See also **rocket motor, solid rocket booster, propellant grain, flame inhibitor**.]

MSS

See **mobile-satellite service**.

MTBF

An acronym for mean time between failures, used in **reliability** assessments. The acronym is also used for mean time before failure, in which case it is equivalent to **MTTF**.

MTTF

An acronym for mean time to failure, used in **reliability** assessments.

MTTR

An acronym for mean time to repair.

MULTI-LAYER INSULATION (MLI)

See **thermal insulation**.

MULTIPACTION

An undesirable effect which can occur in high-power **radio-frequency** (RF) equipment operating in a vacuum. It can occur if the electrical field in a **transmission line** or other component is strong enough to accelerate **electron**s to a velocity at which they are capable of causing the 'secondary emission' of other electrons (e.g. when in collision with an RF electrode or **waveguide** wall). Under the right conditions, the secondary electrons can cause subsequent multiple emissions resulting in an electron avalanche or 'breakdown'. The effect depends on parameters including electrical field strength, RF power level, **frequency** and the dimensions of the transmission line.

Multipaction occurs most readily in coaxial components operating at around 1 GHz, particularly **filters**. Its effects generate **broadband noise** and may, in some equipment, lead to a gas discharge which can cause physical damage. Multipaction is sometimes called 'multipactor' (chiefly in the USA).

MULTIPATH

An undesirable effect caused when a radiated telecommunications **signal** takes more than one path from **transmitter** to **receiver**.

The intended path is always the shortest, 'line-of-sight' path, but in built-up areas particularly, signals may be reflected from solid objects causing them to take a longer path to the receiver [figure M6]. Since both signals travel at the speed of light, the reflected signal reaches the receiver fractionally later than the main signal.

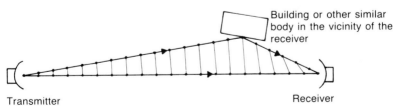

Figure M6 Multipath: signals taking more than one path from a transmitter arrive out of phase at the receiver.

Considering the signal as a simple waveform, multipath constitutes the addition of two waves which combine 'constructively' or 'destructively', to produce a variation in signal level. In the case of a complex TV signal, multipath is manifested on the screen as a faint 'ghost image' or images offset to the right of the main image, since the electron beam which produces the picture scans from left to right across the tube. The secondary signals reach the receiver slightly later than the primary signal, having taken a longer path.

MULTIPLE BEAM

A pattern of several **spot beam**s formed by a **satellite** communications **antenna** with an equal number of **feedhorn**s. This is achieved by placing a cluster of feedhorns at the focus of the antenna reflector. This configuration of multiple feedhorns has two principal uses. If different feedhorns are fed with different power levels and/or different phases, the beam can be shaped to illuminate an irregularly shaped area more efficiently. This can avoid wasting power over areas of empty sea, for instance, and obviate possible interference due to **overspill**, advantages which are not possible with a simple circular or elliptical beam. Alternatively, multiple feedhorns can provide sophisticated communications links between individual beams by switching signals from one beam to another and back in quick succession.
[See also **satellite switching**, **beam-hopping**.]

MULTIPLEXED ANALOGUE COMPONENT (MAC)

A **television** transmission standard developed by the UK Independent Broadcasting Authority (IBA) for use with European DBS (**direct broadcasting by satellite**). The MAC system electronically compresses the luminance and chrominance (brightness and colour) information and transmits them sequentially rather than simultaneously as with existing terrestrial TV standards. The signals are expanded and used to create a picture in the TV receiver. Separating the two information sets means that they cannot interfere and leads to a clearer picture. The main disadvantage is that MAC is not compatible with standard (terrestrial TV) receivers and requires the addition of an adapter or converter.

MULTIPLEXER

A device for combining or separating different signal frequencies; in effect, a multiple **channel filter**. A two-channel multiplexer is called a 'diplexer'. For example, where a number of **channel**s share the **bandwidth** available in a satellite **transponder**, an input multiplexer separates the channel frequencies and routes each **carrier** to its own

amplifier chain. Once amplified, the channels are recombined in an output multiplexer for the return transmission. The individual devices are colloquially known as the input 'mux' and output 'mux', or 'imux' and 'omux' respectively.

MULTIPLEXING

In **telecommunications**, the process of placing more than one **signal** on a **carrier**, **subcarrier** or **channel**. See **frequency division multiplexing (FDM)** and **time division multiplexing (TDM)**.

MULTISTAGE ROCKET

A **launch vehicle** with several **stages**.

MUX

A colloquial abbreviation for **multiplexer**.

N

N (LAUNCH VEHICLE)

A three-stage Japanese **launch vehicle** based on the American **Delta** and built under license in Japan. The propellants used are **liquid oxygen/kerosene** in the first stage, **nitrogen tetroxide/aerozine-50** in the second and a **solid propellant** (**CTPB**) in the third stage and **strap-on** solid boosters. The N launch vehicle has been produced in two variants, the first of which is no longer **operational**: the N-I, which used three strap-ons, had a **payload** capability of 130 kg to **geostationary orbit** (GEO); the N-II, which uses nine strap-ons, can deliver 350 kg to GEO (or about 700 kg to **geostationary transfer orbit** (GTO), the equivalent of a Delta 2910). The N-I made its first flight, from the launch site on **Tanegashima** Island off the south coast of mainland Japan, in September 1975, and its seventh and last in September 1982; the N-II made its first flight in February 1981.
[See also **H (launch vehicle).**]

NACA

See **NASA**.

NARROWCASTING

The dissemination or transmission of a broadcast-type signal to a closely defined audience with a particular interest, as opposed to **broadcasting** which does not imply application to a particular audience.

NASA

An acronym for National Aeronautics and Space Administration, a body formed in October 1958 to coordinate United States research and development into aeronautics, **space science** and **space technology**. NASA was based on the structure of its predecessor, the National Advisory Committee for Aeronautics (NACA). NASA's main establishments are as follows:

Headquarters, Washington, DC (includes six programme offices: Office of Aeronautics and Space Technology (OAST); Office of Space Flight; Office of Space Science and Applications (OSSA); Office of Space Station; Office of Space Tracking and Data Systems: Office of Commercial Programs);

Ames Research Center†, Mountain View, California (aeronautics research, computer science, flight simulation, biomedical research, etc);

Dryden Flight Research Center†, **Edwards Air Force Base,** California (flight research [see **X-1, X-15, Enterprise**]);

Goddard Space Flight Center (GSFC), Greenbelt, Maryland (meteorological, communications and scientific satellites);

Jet Propulsion Laboratory (JPL), Pasadena, California (interplanetary spacecraft);

Johnson Space Center (JSC), Houston, Texas (manned spaceflight, **astronaut** training and mission control);

Kennedy Space Center, Florida (launch facilities);

Langley Research Center, Hampton, Virginia (advanced aeronautical and **materials** research);

Lewis Research Center, Cleveland, Ohio (space **power** and **propulsion systems**, etc);

Marshall Space Flight Center, Huntsville, Alabama (**launch vehicle** development, space science, etc);

Stennis Space Center (formerly National Space Technology Laboratories), Mississippi (static tests of large propulsion systems);

Wallops Flight Facility, Wallops Island, Virginia (launch facilities for **sounding rockets**, scientific balloons, etc);

White Sands Test Facility, New Mexico (**propellant** and **rocket engine** testing, etc).

NASDA

An acronym for NAtional Space Development Agency of Japan, a body responsible for the development of Japanese space technology and the implementation of its applications (established in October 1969). NASDA facilities include the Tsukuba Space Center (research and development, **tracking** and control), Kakuda Propulsion Center (**propulsion systems**) and the Tanegashima Space Center (**launch site**). [See also **Tanegashima**.]

---

†Ames and Dryden were merged in 1981 to form the 'NASA Ames Dryden Flight Research Facility', referred to as 'Ames-Moffet' and 'Ames-Dryden'.

NATIONAL SPACE TECHNOLOGY LABORATORIES

See **NASA**.

NAVIGATION SATELLITE

A satellite, usually part of a system providing global coverage, which gives accurate position information to mobile terrestrial **receiver**s. The definition includes military and civilian systems serving maritime, aeronautical and land mobiles. One early example is the COSPAS-SARSAT programme (Kosmicheskaya Systyema Poiska Avariynych Sudov—Search And Rescue Satellite-Aided Tracking). The system, declared **operational** in 1984, comprises a number of satellites in low, near-polar orbits which receive transmissions from emergency location beacons (known on aircraft as ELTs (emergency location transmitters) and EPIRBs (emergency position-indicating radio beacons) on ships).

Another example is the Navstar GPS (global positioning system) which, when completed, will comprise a **constellation** of 18 satellites (plus three **in-orbit spare**s) in six different 12-hour orbits inclined at 55 degrees to the equatorial plane [see figure C3]. Although the Navstar system was initiated for the US military services, it can also be used for commercial activities such as business aviation and geological surveys, albeit at lower positional accuracies.
[See also **mobile-satellite service (MSS)**.]

NEGATIVE MARGIN

See **margin**.

NERVA

An acronym for nuclear engine for rocket vehicle application, the name given to a nuclear-powered rocket development programme of the 1960s, which was cancelled largely due to lack of funds. See **nuclear propulsion**.

NEWTONIAN TELESCOPE

A type of astronomical reflecting telescope in which light is reflected from a parabolic primary mirror onto a plane secondary and through an aperture in the side of the telescope to an eyepiece [figure N1]. Named after Sir Isaac Newton (1642–1727), the English scientist and mathematician.
[See also **space telescope, Schmidt telescope, Cassegrain reflector, Gregorian reflector, Coudé focus**.]

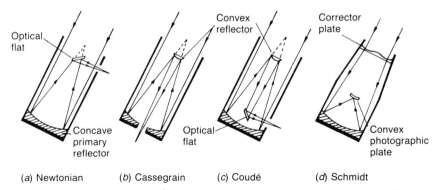

(*a*) Newtonian   (*b*) Cassegrain   (*c*) Coudé   (*d*) Schmidt

Figure N1 Telescope optics [see **Newtonian telescope, Cassegrain reflector, Coudé focus, Schmidt telescope**].

NICKEL–CADMIUM CELL

A commonly used cell in a spacecraft **battery**.

NICKEL–HYDROGEN CELL

A commonly used cell in a spacecraft **battery**.

NITROGEN TETROXIDE (N_2O_4)

A storable liquid **oxidiser** used in **rocket engines**. See **liquid propellant**.

NOAA

An acronym for National Oceanographic and Atmospheric Agency, a US Government body that operates weather satellites. See **meteorological satellite**.

NOISE

An undesirable electrical disturbance which, if sufficiently severe, can mask the **signal** in a communications system [see **signal-to-noise ratio, carrier-to-noise ratio**]. Noise has a number of origins: the equipment itself [see **thermal noise, shot noise**]; other electrical devices such as motors and ignition systems; natural terrestrial or atmospheric noise such as that from electrical storms (significant only below about 30 MHz); and extraterrestrial noise, particularly from the Sun. Noise caused by natural electrical discharges in the atmosphere is sometimes called 'static'.

[See also **Boltzmann's constant**.]

NOISE FACTOR (F)

See **noise figure (NF)**.

NOISE FIGURE (NF)

A measurement of the **noise** contribution of a device (e.g. an **amplifier**) expressed relative to a theoretical noise-free amplifier at a reference temperature. The noise figure of a device is its 'noise factor' expressed in dBW [see **decibel**]. Noise factor, F, can be calculated using the following expression:

$$F = P_N/GkT_aB$$

where P_N is the **noise power** at the output, G is the **gain** and kT_aB is the noise power of the theoretical source resistance at ambient temperature (where k is **Boltzmann's constant**, T_a is the normal ambient temperature of the source resistor, taken to be 290 K, and B is the effective noise **bandwidth** of the system in Hz).

If no noise is contributed by the device, the noise factor is 1 and the noise figure is 0 dBW.

NOISE POWER

A quantitative measure of **noise** in a communications system (units dBW), analogous to **carrier** power; the quantity used to calculate the **carrier-to-noise (power) ratio**.

Noise power, P_N, is given by the expression

$$P_N = kTB$$

where k is **Boltzmann's constant**, T is the noise temperature of the source and B is the effective noise **bandwidth** of the system. See **noise temperature**.

[See also **noise figure**.]

NOISE POWER DENSITY (NPD; N_0)

In **telecommunications**, a quantitative measure of the level of **noise** in every 1 Hz of **bandwidth** (units dBW Hz^{-1}). NPD is given by the expression

$$N_0 = P_N/B = kT$$

where P_N is the noise power (dBW), B is the bandwidth (dBHz), k is **Boltzmann's constant** and T is the **absolute temperature** (dBK).

[See also **noise temperature, noise figure, decibel**.]

NOISE TEMPERATURE

In **telecommunications**, a concept which relates the **noise power** at the source (or other reference point) of a system to the temperature of a reference resistance from which the same 'thermal power' is available. Put another way, the noise temperature is the temperature at which the reference produces an equivalent noise power to that produced by the system under consideration, over the same **frequency** range (or **bandwidth**). It is given by the expression

$$T = P_N/kB$$

where P_N is the noise power (dBW), k is **Boltzmann's constant** and B is the bandwidth (dBHz). The units of noise temperature are dBK (in the same way that **power** is referenced to the watt in dBW, temperature is referenced to the kelvin in dBK).

For an orbiting **satellite** 'looking at' the Earth, the noise temperature is the sum of the **thermal noise** of the **receiver** equipment and the noise due to the Earth. The Earth's temperature from space is an average of 290 K, so its noise temperature is 24.6 dBK ($10\log_{10}290$).

The Sun is a strong variable noise source with a noise temperature of about 10^6 K at 30 MHz and at least 10^4 K at 10 GHz under quiet Sun conditions—large increases occur at times of greater solar activity. [See also **noise figure**.]

NOMINAL

(i) 'Theoretical' or 'intended' (a 'nominal value': e.g. 'nominal **capacity**', 'nominal **orbit**', etc [see **nominal orbital position**]). Usage, e.g. 'its deviation is nominally zero'.

(ii) Jargon: used to describe the performance of a device, subsystem, etc, which is 'within specification' (i.e. operating as intended). Usage, e.g. 'power subsystem is nominal'.

NOMINAL ORBITAL POSITION

The intended, average **orbital position** of a **satellite** in **geostationary orbit**.

An allocated orbital position cannot be maintained precisely, due to the spacecraft's susceptibility to a variety of perturbing forces, chiefly **gravitational perturbation**s and the **solar wind**. The gravitational attraction of the Sun and Moon, for example, tend to 'pull' the satellite from its nominal position while the solar wind 'pushes'. The mass and surface area of the spacecraft and the relative positions of the astronomical bodies govern the magnitude of these effects and vary the amount of **thruster** fuel utilised for **station keeping**. [See also **orbit**, **orbital spacing**, **orbital control**.]

NON-OPERATIONAL

The opposite of 'operational'. See **operational**.

NORAD

An acronym for NORth American Air Defence, a military agency in the USA which, amongst other things, tracks and monitors **spacecraft** and other objects in Earth **orbit**. For example, the NORAD catalogue for 30 July 1987 showed 6746 man-made objects in orbit, 5108 of which were classed as debris.
[See also **orbital debris**.]

NORTH–SOUTH STATION KEEPING

See **orbital control**.

NORTHERN COSMODROME

A Soviet **launch site** near the town of Plesetsk (at approximately 63°N, 40°E) used predominately for the launch of satellites into high-**inclination** orbits. The Northern Cosmodrome is largely military in nature, the Soviet equivalent of the US **Vandenberg Air Force Base**.
[See also **Baikonur Cosmodrome**.]

NOSE CONE

A colloquial term for the forward end of a **rocket**; the section which generally contains a **payload**. The pointed tip of a rocket booster is technically known as the '**ogive** section'. The section containing the payload is referred to generally as a **fairing** and specifically as a 'payload shroud' or 'launch shroud'. See **shroud**.

NOSE FAIRING

An alternative term for **nose cone**.
[See also **ogive**.]

NOZZLE

The part of a **rocket engine**, **rocket motor** or **thruster** which converts the thermal energy of the **combustion** gases into the kinetic energy of the exhaust **plume** [see figure N2 and frontispiece]. Known more fully as an 'exhaust nozzle' or 'expansion nozzle'.

A typical nozzle comprises two sections: a convergent section and a divergent exit cone. The narrowest part of the nozzle, called the throat, is designed to maintain the required pressure within the **combustion chamber** and regulate the outflow of combustion gases. The gases move naturally from the high pressure of the chamber to the vacuum of space, expanding and accelerating rapidly as they leave the chamber. The throat is the region of transition from subsonic to supersonic flow. The exit cone controls the expansion of the exhaust plume [see **expansion ratio**].

Figure N2 Space Shuttle launch showing the **nozzles** of the **Space Shuttle main engines** (SSMEs) and a **solid rocket booster** (SRB). The Shuttle's control surfaces are clearly visible [see **elevon**] as are the attachment struts connecting the **orbiter** and the **external tank**. [NASA]

The convergent-divergent shape of the nozzle is due to the variation in the rates of change of the velocity and specific volume (volume per unit mass) of the combustion gases: since the rate of increase of specific volume is at first less than, but finally greater than, the rate of increase of velocity, the nozzle must first converge and then diverge. [See also **exhaust velocity, flow separation, regenerative cooling, mach number, plug nozzle.**]

NTO

An abbreviation for **nitrogen tetroxide (N_2O_4)**.

NUCLEAR PROPULSION

A form of rocket propulsion which uses the heat from a nuclear reactor to vaporise a **propellant** (typically liquid hydrogen), the resultant gas being ejected from a **nozzle** to produce **thrust** in the same way as any chemical propulsion system. Although there are currently no operational systems, nuclear propulsion remains an attractive prospect for the future since it promises a very high **specific impulse** (around 1000 s; cf 400 s for **liquid hydrogen/liquid oxygen**).
[See also **NERVA**, **chemical propulsion**, **electric propulsion**.]

NUTATION

A periodic variation in the **precession** of a spinning body (e.g. a planet, **satellite** or **gyroscope**); a 'nodding' motion superimposed on the precession of the spin axis [see figure N3].
[See also **nutation damper**.]

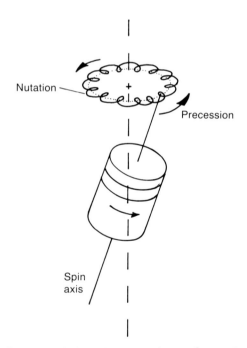

Figure N3 Spin axis **precession** and **nutation**.

NUTATION DAMPER

A device designed to reduce the **nutation**, or 'nodding' motion, of a spinning spacecraft.

Most important for **spin-stabilised** spacecraft, but also used for

three-axis-stabilised satellites which are spin stabilised during their **transfer orbit**s. Active nutation damping is realised using **reaction control thruster**s: since the spacecraft is spinning, most thruster firings have to be made in **pulse mode**, pulsing each time the spacecraft's rotation brings the thruster into alignment with the desired **thrust vector**. Passive damping involves a device containing either a viscous liquid in a sealed circuit or a steel ball in a gas-filled tube: the energy of nutation is converted to heat in the working fluid, typically mercury in the first instance and neon in the second, and the motion is damped out.

[See also **DANDE**.]

O

OBSERVATORY

Usually an institution or building designed and equipped to perform observations of astronomical or meteorological phenomena. The term is also applied to spacecraft designed to conduct astronomical observations from space (e.g. Orbiting Solar Observatory, OSO; **space telescope**).

OFFSET FEED

An **antenna** system in which the **feedhorn** is placed away from the principal axis of the antenna reflector: it is 'offset' from the centre. The antenna is described as 'offset-fed'. In most designs the horn is also placed outside the **beam** formed by the reflector, to eliminate the 'blockage' and diffraction effects caused by a horn which intercepts the beam. An antenna with a feedhorn on the principal axis is known as 'centre-fed' [see figure A2].

Both spacecraft and **earth station** antennas can be either offset-fed or centre-fed.

OGIVE

A section of a **launch vehicle** fairing or tank shaped like a pointed (Gothic) arch: usually the forward end (e.g. the forward end of the Space Shuttle **external tank** [figure E6]).
[See also **fairing**.]

OMNI

Colloquial abbreviation for **omnidirectional antenna**.

OMNIDIRECTIONAL ANTENNA

Theoretically, an **antenna** with coverage in all directions (in three-dimensional space), commonly used on spacecraft as a **telemetry, tracking and command** antenna. Often abbreviated to 'omni'.

In practice the 'antenna' may consist of two or more separate radiating elements, mounted at opposite ends of the spacecraft, to provide two (or more) overlapping hemispheres of coverage. Because of this, and the general difficulties in providing a perfect **antenna radiation pattern**, the omni cannot be described as **isotropic** since the pattern is not equal in all directions. [See figure D2 for an example of a deployable omni and figure G3 for one mounted on a satellite's **feedhorn** tower.]
[See also **isotropic antenna**.]

OMS

See **orbital manoeuvring system**.

OMT

See **orthomode transducer**.

OMUX

A colloquial abbreviation for output multiplexer. See **multiplexer**.

OMV

See **orbital manoeuvring vehicle**.

ON-STATION

The condition of a satellite in **geostationary orbit** located at its **nominal orbital position**.

OPERATIONAL

Generally: 'capable of, or actually involved in operations'. For example, a new **launch vehicle** is typically deemed 'operational' following the successful completion of a number of test launches (four in the case of both the European **Ariane** and the American **Space Shuttle**). A **satellite** generally becomes operational when it has reached its allocated **orbital position** and successfully completed a series of **performance** tests.

In some cases **satellite communications** systems are considered 'operational' when there are sufficient resources (i.e. satellites and satellite **channels**) to offer a continuous and guaranteed service: usually this requires at least two satellites in orbit for a particular system, providing **redundancy** should one of them fail. While there was only one satellite in orbit, such a system would be considered **pre-operational**.

[See also **in-orbit spare**, **ground spare**, **lifetime**.]

OPERATIONAL LIFETIME

See **lifetime**.

OPF

See **orbiter processing facility**.

OPTICAL FIBRE

Fibres of glass, usually manufactured in continuous lengths of several kilometres and packed in bundles to form a cable, used for the transmission of signals modulated onto a laser beam. The fibres are generally made from silica glass (with the chemical formula SiO_2) with dopants such as germania (GeO_2) or fluorine (F) added to produce small changes in the refractive index. The fibre core is doped to a higher index than the surrounding cladding to confine the light in the core by total internal reflection.

Optical fibre offers competition to **communications satellite**s over the long-distance trunk routes, due in part to its high transmission capacity and relatively low cost. Its advantages over **coaxial cable** include higher capacity, larger **repeater** spacings and immunity from electromagnetic **interference**.

OPTICAL GYROSCOPE

See **laser gyro**.

OPTICAL SOLAR REFLECTOR (OSR)

A material which reflects the Sun's visible spectrum. This term is often used synonymously with **second-surface mirror** (SSM), but is less definitive since it includes other solar reflectors, such as white paint, silver-coated silica sheets, etc [see figure S2].

[See also **solar spectrum**.]

ORBIT

The path followed by an object (**spacecraft** or **planetary body**) in its motion around a **celestial body**; to follow such a path.

Two periods of revolution can be defined for an orbiting body, dependent on the **frame of reference**: the orbital period with respect to the stars is known as the 'sidereal period' and the period with respect to the celestial body is called the 'synodic period'. Consider, for example, a satellite in an equatorial, **low Earth orbit** and an observer fixed in space; both the Earth and the satellite revolve from west to east. The observer, ignoring the Earth entirely, notes the time the satellite takes to describe a complete circuit against the background of stars—this is its sidereal period of revolution. In this time, however, the Earth has rotated to a new position and the line of longitude that was directly beneath the satellite at the beginning of the sidereal period is now ahead of it. The observer notes the time the satellite takes to catch up with the original **sub-satellite point** and adds this to the sidereal period —this is the synodic period of revolution. For any orbit the synodic period can be defined as the time between two passes over the reference longitude.

In the early days of spaceflight, the practice arose of referring to the synodic periods as 'revolutions' and the sidereal periods as 'orbits' (e.g. Gemini 7 completed 206 revolutions and 220 orbits during its 14-day mission).

[See also **geostationary orbit, geosynchronous orbit, helio-synchronous orbit, Molniya orbit, tundra orbit, transfer orbit, Hohmann transfer orbit, drift orbit, prograde orbit, retrograde orbit, graveyard orbit, orbital position, orbital parameters, orbital debris, perturbations, apogee, perigee, trajectory.**]

ORBITAL CONTROL

The ability to maintain or change a spacecraft's **orbital parameters**; the subsystem or process by which this control is effected.

Most spacecraft required to orbit a **celestial body** have an orbital control subsystem, since they need to counter the forces which combine to perturb that orbit [see **perturbations**]. The system usually comprises a set of **reaction control thruster**s.

The satellite in **geostationary orbit** (GEO) is, however, a special case, since it is allocated an **orbital position** measured in degrees of longitude, which must be maintained to limit the possibility of **interference** with other services using GEO, to allow the use of fixed ground **antenna**s and to avoid collisions. This is referred to as 'station keeping': the two main components of orbital control are east—west station keeping and north—south station keeping. The former relates to a drift in longitude along the geostationary arc away from the '**nominal orbital position**', the latter to a drift north or south of the equatorial plane [see **free-drift strategy**]. Orbital control is usually combined with **attitude control** in a common subsystem, the 'attitude and orbital control system' (AOCS).

The orbital parameters of a spacecraft can be determined using **star sensor**s and an onboard computer, or by the measurement of distance and direction from the ground station [see **tracking**].
[See also **sun sensor, earth sensor, equilibrium point**.]

ORBITAL DEBRIS

A blanket term for any man-made artefact discarded, or accidentally produced, in **orbit** (chiefly around the Earth). This includes satellites which have reached **nd of life**, spent **launch vehicle** stages, hardware accidentally released by **astronaut**s and the remnants of **spacecraft** which have exploded.

Objects about 10 cm in diameter can be tracked in **low Earth orbit** (LEO) and those greater than 1 m in diameter can be detected in **geostationary orbit** (GEO). They are monitored by the North American Air Defense (**NORAD**) organisation, which shows the maximum density of material at 850 km. The probability of collision with orbital debris has been estimated: for a Space Shuttle **orbiter** on a typical 7-day mission in LEO it is about 4×10^{-6}; for a satellite in GEO it lies between 10^{-6} and 10^{-5} in one year (about a one in a million chance).

The best way to reduce the probability of collision is to prevent the accumulation of debris, which for geostationary satellites is achieved by removing spacecraft from orbit at the end of their lives [see **graveyard orbit**].

ORBITAL DECAY

The gradual reduction in the altitude of an **orbit** due to **atmospheric drag**. If it remains uncorrected, the decay of an orbit results eventually in an uncontrolled **re-entry**.
[See also **drag compensation**.]

ORBITAL ELEMENTS

See **orbital parameters**.

ORBITAL INCLINATION

The angle between the plane of an **orbit** and the equatorial plane of the orbited body.
[See also **plane change, orbital control**.]

ORBITAL INJECTION

See **injection**.

ORBITAL MANOEUVRING SYSTEM (OMS)

A **propulsion system** on the American **Space Shuttle** used for changing orbits, orbital **injection**, **rendezvous** and **de-orbit**. It comprises two 26 700 N (6000 lb) engines mounted in two removable 'OMS pods' (commonly pronounced 'ohms pods') on either side of the **orbiter**'s tail [figure N2].

The pods also house part of the orbiter's **reaction control system** (RCS): twenty-four 3870 N (870 lb) primary thrusters and four 111.2 N (25 lb) vernier thrusters (an additional 14 primary and 2 vernier thrusters are located in a module near the orbiter's nose). All OMS and RCS thrusters use the **hypergolic** propellants monomethyl hydrazine (**MMH**) and **nitrogen tetroxide**.

ORBITAL MANOEUVRING VEHICLE (OMV)

The concept of a spacecraft designed to reposition **payload**s in **orbit**; more colloquially known as a 'space tug'. In its ultimate form such a vehicle would be re-usable, by being refuelled, and would remain in orbit throughout its **operational** life, rather than be returned to Earth. [See also **orbital transfer vehicle**.]

ORBITAL PARAMETERS

The parameters which define a spacecraft **orbit** (namely **semi-major axis**, **eccentricity**, inclination, argument of perigee, right ascension of ascending node, true anomaly).
[See also **nominal orbital position**, **orbital inclination**, **apogee**, **perigee**, **drift**.]

ORBITAL PERIOD

Of a spacecraft, in general taken to be the **sidereal period** (as opposed to the **synodic period**). See **orbit**.

ORBITAL POSITION

The position of a **satellite** in **geostationary orbit** referred to the Earth's longitude at the **sub-satellite point** and measured in degrees east or west of the Greenwich Meridian.

The allocation of orbital positions (as well as frequencies) is coordinated through the International Telecommunications Union (**ITU**). The concept of an orbital position is not relevant to non-geostationary spacecraft, since they do not remain over a particular line of longitude.
[See also **nominal orbital position**, **orbit**, **orbital spacing**, **orbital control**.]

ORBITAL RELOCATION

The movement of a spacecraft in **geostationary orbit** from one **orbital position** to another. This occurs typically when a **communications satellite** is required to provide coverage to an area different from that which was originally intended. The manoeuvre may be performed using the satellite's orbital control **thrusters** or the spacecraft may be allowed to drift until the required position is reached, dependent upon the relative positions. See **equilibrium points**.
[See also **coverage area, service area**.]

ORBITAL SLOT

A colloquial term for **orbital position**.

ORBITAL SPACING

The separation of satellites in **geostationary orbit** measured in degrees of longitude.

Although the precise separation values vary around the **orbit**, the general rule is two degrees between satellites operating at **Ku-band**, three degrees for **C-band** and six degrees for **DBS** satellites.
[See also **orbital position**.]

ORBITAL TRACK

The projection of a satellite's **orbit** on the surface of a **planetary body**, otherwise known as a '**ground track**'; the locus of **sub-satellite points** drawn out as the satellite orbits the body.

ORBITAL TRANSFER VEHICLE (OTV)

The concept of a spacecraft designed to transfer a **payload** from one **orbit** to another, typically from a **low Earth orbit** (LEO) to a higher altitude orbit (e.g. **geostationary orbit** (GEO)), or from an **equatorial orbit** to a **polar orbit**.

In its ultimate form such a vehicle would be re-usable, by being refuelled, and would remain in orbit throughout its **operational** life, rather than be returned to Earth. The closest contemporary vehicle is the expendable **perigee kick motor**, a propulsive stage designed to transfer a spacecraft from LEO to GEO.
[See also **plane change, orbital manoeuvring vehicle**.]

ORBITER

Another name for the American **Space Shuttle** [figure O1]; the part of the **space transportation system** which attains **orbit** and operates as a **spacecraft** in orbit, as opposed to being simply part of a **launch vehicle** [see **external tank, solid rocket booster**].

The Shuttle orbiter is 37.18 m long, 17.39 m high and has a 23.77 m wingspan. It is designed to operate in **low Earth orbit**, i.e. between altitudes of 185 and 1110 km (115–690 miles).

[See also **payload bay, flight deck, Space Shuttle main engine (SSME), auxiliary power unit (APU), orbital manoeuvring system (OMS), remote manipulator system (RMS), thermal protection system (TPS)**.]

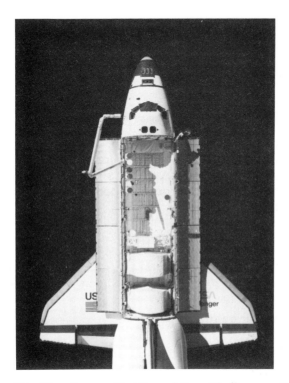

Figure O1 The Space Shuttle **orbiter** Challenger. Note the **sunshields** protecting two **communications satellites** in the **payload bay** and the row of **getaway special** canisters on the left sill. Also visible is the deployed **remote manipulator system** (RMS) and, on the orbiter's nose, the **thruster** ports of the reaction control system [see **orbital manoeuvring system**]. [NASA]

ORBITER PROCESSING FACILITY (OPF)

The building at **Kennedy Space Center** in which American Space Shuttle **orbiter**s are prepared for flight; the Shuttle equivalent of an aircraft hanger. It is divided into two 'high bays', each capable of accommodating a Shuttle orbiter in a horizontal attitude, and an interconnecting 'low bay' containing ancilliary equipment and offices, etc. **Payload**s which require horizontal **integration** are loaded in the OPF; those which can be integrated with the orbiter in a vertical attitude are loaded at the **launch pad** using a **payload canister**. The orbiter is towed from the OPF to the **vehicle assembly building** (VAB) for assembly with the **external tank** (ET) and **solid rocket booster**s (SRBs). [See also **processing**.]

ORTHOGONAL AXIS

A perpendicular axis. Commonly used to refer to axes of radiation **polarisation**: for example, vertical and horizontal polarisations are 'orthogonal'. The major and minor axes of an **elliptical antenna** are 'orthogonal'.
[See also **spacecraft axes**.]

ORTHOMODE TRANSDUCER (OMT)

A component of a **waveguide** circuit, known more fully as an orthogonal-mode transducer, that separates or combines two orthogonally polarised signals. An OMT allows the same **antenna** and **feedhorn**(s) to be used with transmitted and received signals of opposite polarity.

OSR

See **optical solar reflector**.

OTV

See **orbital transfer vehicle**.

OUTDOOR UNIT

See **head-end unit**.

OUTGASSING

The release of a gas from a material when it is exposed to an ambient pressure lower than the vapour pressure of the gas. Many materials (resins, adhesives and even some metals) evaporate in a vacuum. Since there is no force available to carry the particles away from a spacecraft, they remain in a cloud around it or condense onto its colder surfaces. Spacecraft **materials** are chosen to limit outgassing, since the outgassed molecules can degrade the optical surfaces of **sensor**s, **solar cell**s and **radiator**s, and even cause shorting of electrical circuits and promote 'corona' or electrical discharge.

OUTPUT

(i) The exit point or path through which information or energy is removed from a device or system (refers to the hardware).

(ii) The information or energy (a **signal**, **carrier**, **RF wave**, etc) passed out of an electronic or **microwave** circuit or component.

[See also **output section**, **output filter**, **data**.]

OUTPUT FILTER

A **filter** at the output end of a **transmit chain** in a communications **transponder** which confines the signal to its appointed **bandwidth**. The typical output filter may comprise more than one individual filtering device, but its major constituent is known generically as a **band-pass filter**.
[See also **input filter**.]

OUTPUT SECTION

The section of a spacecraft **communications payload** from the signal's input to the **travelling wave tube amplifier** to its output to the transmit antenna. The division of payload components into an **input section** and on output section is an alternative to their classification as part of a **receive chain** or a **transmit chain** [see figure C2].

OVER-DESIGN

In space technology the term usually refers to an item which has been designed to be stronger or generally more capable than necessary (e.g. a structural item designed to withstand a much greater structural **load** than it can be expected to experience is 'over-designed', as is a **sensor** designed to operate at much greater accuracies than will ever be required). In general, the practice of over-design increases the cost, and in some cases the mass, of the item, and is therefore best avoided.

OVEREXPANDED

See **expansion ratio**.

OVERSPILL

The unavoidable tendency of a **communication satellite**'s antenna **beam** to cover more than the intended **service area**. Sometimes called spillover, but this term is more often used in connection with **antenna radiation pattern**s—see **spillover**.

Overspill results from the necessity to design satellite **footprint**s to provide sufficient **power flux density** at the edge of the service area even under the worst weather conditions (when the **signal** is most highly attenuated). This built-in **margin** means that, under 'clear sky' conditions, the effective **coverage area** is increased and the footprint spills over into adjacent states or countries. If the degree of overspill is too large, **interference** with other services may result.
[See also **attenuation**.]

OXIDISER

The component of a rocket **propellant** which allows the **combustion** of a **fuel** in the absence of atmospheric oxygen (e.g. **liquid oxygen, hydrogen peroxide** (H_2O_2), **nitrogen tetroxide** (N_2O_4), **ammonium perchlorate** (NH_4ClO_4)).
[See also **solid propellant, liquid propellant**.]

P

PAD

Engineering slang for an **attenuator** (e.g. 'a 3 dB pad'). Also used as a verb: 'to pad a **transmission line**', etc.
[See also **launch pad**.]

PAHT

An acronym for power-augmented hydrazine thruster. See **hiphet thruster**.

PALLET

A portable U-shaped support and carrying structure upon which **payload**s can be mounted prior to loading into the **payload bay** of the American **Space Shuttle** [see figure P1]. The use of pallets decreases the time required for Shuttle **payload integration** and means that complex payloads can be assembled on a pallet and tested independent of Shuttle **processing**. They are available as single 2.9 m long pallets or 1.4 m 'half-pallets', which can be combined to form 'pallet trains' dependent on the size of the payload. Although they have many potential applications, the pallets are sometimes referred to as '**Spacelab** pallets', since they were originally designed for use with Spacelab **habitable module**s (as carriers of instruments requiring direct exposure to space). Pallets are intended primarily to remain within the payload bay, but it would be possible to release them as orbiting spacecraft in their own right (so-called 'free-flying pallets').
[See also **cradle, airborne support equipment**.]

PAM

See **payload assist module**.

Figure P1 A **pallet** carrying scientific equipment prepared for flight on the American **Space Shuttle**. [ESA]

PARABOLA

A conic section formed by the intersection of a cone by a plane parallel to its side.

PARABOLIC ANTENNA

An **antenna** with a parabolic cross section; an antenna shaped like a **paraboloid** (i.e. 'paraboloidal').

PARABOLIC TRAJECTORY

See **ballistic trajectory**.

PARABOLOID

A geometric, three-dimensional surface which has parabolic cross sections in two of three orthogonal planes and either elliptical or hyperbolic sections in the third. For example, the main reflector of a **Cassegrain reflector**.
[See also **Gregorian reflector**.]

PARALLEL TUBES

Two or more **travelling wave tube**s (TWTs) coupled together in parallel
to provide a higher power than that available or conveniently obtained
from a single TWT. Parallelling two similar tubes does not produce
twice the power of an individual tube due to losses suffered in the
combination process, but if one tube should fail the system can still
operate, albeit on reduced power.

PARAMETRIC AMPLIFIER

A type of **low-noise amplifier** which has one oscillator circuit tuned to
the receive frequency and another 'pumping' oscillator at a different
frequency. The pumping oscillator periodically varies the parameters
of the primary circuit (capacitance or inductance), thereby transferring
energy to and thus amplifying the input signal.

PARAMP

Colloquial abbreviation for **parametric amplifier**.

PARASITIC STATION ACQUISITION

A technique used to deliver satellites with liquid **bipropellant
propulsion system**s to **geostationary orbit** with a number of separate
thruster firings, as opposed to the single '**burn**' produced by a **solid
rocket motor**. This philosophy allows the engine to be calibrated on the
first burn, thus increasing the accuracy of subsequent burns. By
varying the timing and magnitude of the burns to adjust the periods of
intermediate **transfer orbit**s, the final burn can be performed at the
desired **on-station** longitude, injecting the satellite directly into its final
orbit at the correct position. This technique can save a significant
amount of **station-keeping** fuel and increase the satellite's **lifetime**.
[See also **mid-course correction**.]

PARKING ORBIT

A term used for a **low Earth orbit** (or similar orbit around any other
planetary body) when used as a temporary 'holding position' for a
spacecraft, before it is injected into another **orbit** or **trajectory**.
[See also **geostationary transfer orbit, trans-lunar injection**.]

PARSEC

The distance at which an object displays a parallax of 1 second of arc
across a baseline of 1 **astronomical unit** (i.e. 3.086×10^{16} m); a unit of
distance used in astronomy.
  Derivation: measurements taken of relative star positions, viewed

from either side of the Earth's orbit about the Sun (i.e. six months apart), show a movement of the nearer stars, against the distant background stars, known as parallax. The observation baseline is two astronomical units; if observations are adjusted for a baseline of 1 AU, a parallax of 1 second indicates a distance of 1 parsec (from 'PARallax SECond'). One parsec is equivalent to about 3.262 **light years**.

PASSIVE

A term applied to any device or system which is receptive or responsive to external stimuli in a pre-arranged manner, but makes no individual contribution to the stimulus; the opposite of **active**. For example, a passive device in a **communications system** contains no power source to augment the system's output power; all power is derived from the input **signal** (e.g. a **filter**, **attenuator** or **hybrid**). Other examples of passive components are the **surface coatings**, **thermal insulation**, etc, applied to spacecraft as part of a **thermal control subsystem**.

PASSIVE COMMUNICATIONS SATELLITE

A **communications satellite** which has no **active** components; e.g. Echo 1 and 2, two spherical metallised balloons (30 m and 41 m in diameter) placed in **low Earth orbit** in August 1960 and January 1964 respectively. The balloons simply reflected the radio signals beamed at them.

PAYLOAD

Any equipment or cargo carried by a **spacecraft** or **launch vehicle**, or mounted on an orbiting **space platform**. The term may also include the crew of a **manned spacecraft**.
[See also **communications payload**, **payload module**.]

PAYLOAD ASSIST MODULE (PAM)

A **solid rocket motor** designed for use with the American **Space Shuttle**, **Delta** and **Titan** launch vehicles as a **perigee kick motor** or 'perigee stage' to inject satellites from **low Earth orbit** into a **geostationary transfer orbit**. Also known as a spinning solid upper stage (SSUS), since it uses **solid propellant** and is capable of spin-stabilising its **payload** during the transfer orbit. The PAM is available in two main versions: PAM-A is designed for payloads in the **Atlas—Centaur** class (about 1000 kg to **geostationary orbit** and PAM-D for smaller payloads suited to the Delta (about 550 kg to GEO). A variant, the PAM-DII, has a payload capability similar to that of the PAM-A [see figures P2 and P3].
[See also **upper stage**, **spin table**, **spin-stabilised**.]

Figure P2 A **spin-stabilised** communications satellite being deployed from the American **Space Shuttle**. The satellite is mounted on a **payload assist module** (PAM). [NASA]

PAYLOAD BAY

The 'cargo hold' of the American Space Shuttle **orbiter**. It is 18.28 m long and 4.57 m in diameter and can accommodate either one large or several smaller **payload**s [see figures P4 and O1]: up to five Spacelab **pallet**s, or the equivalent in pallets and **habitable module**s, or up to six average-sized **satellite**s mounted on individual support **cradle**s, can be carried.
[See also **remote manipulator system (RMS)**, **Spacelab**.]

PAYLOAD CANISTER

A container about the same size as the **payload bay** of the American **Space Shuttle** used to transfer **payload**s to the **launch pad** under 'clean-room' conditions [see figure T1].
[See also **rotating service structure (RSS)**.]

PAYLOAD CHANGEOUT ROOM

See **rotating service structure (RSS)**.

Figure P3 A **three-axis-stabilised** communications satellite being deployed from the American **Space Shuttle**. The satellite is mounted on a **payload assist module** (PAM). [NASA]

PAYLOAD ENVELOPE

The dimensional constraint on the volume available for a **payload** within a **launch vehicle** shroud [see figure P5]. The envelope constrains the size and shape of a spacecraft and its various appendages, such as **antenna**s and **solar array**s, with the result that most of them have to be folded against the side of the spacecraft for launch and **deploy**ed once in **orbit**.
[See also **shroud**.]

PAYLOAD FRACTION

See **mass ratio**.

PAYLOAD INTEGRATION

(i) The loading of one or more **spacecraft** onto a **launch vehicle**.

(ii) The assembly of the components of a spacecraft **payload** (e.g. **communications payload**.).

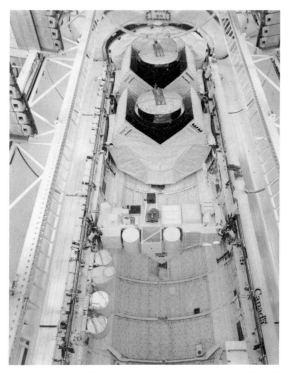

Figure P4 The **payload bay** of the American **Space Shuttle** prior to the launch of the STS 41-B mission carrying (from top) the Palapa B2 and Westar VI communications satellites, the SPAS-01 shuttle pallet satellite and a number of **getaway special** canisters. Compare the open **sunshield**s with figure M1. Note also the stowed **remote manipulator system** on the right-hand sill. [NASA]

(iii) Sometimes the mating of a spacecraft's **payload module** with its **service module**, but this is usually called **spacecraft integration**.

PAYLOAD MASS RATIO

See **mass ratio**.

PAYLOAD MODULE

A self-contained section of a modular spacecraft containing the **payload** (e.g. the **communications payload** of a **communications satellite**, otherwise known as the communications **module**) [see figures M2 and M3].

For the satellite the term may include the communications **antenna**s or they may be contained within a separate **antenna module**, depending on the design. The other main modules are the **service module** and **propulsion module**. The chief advantage of the modular design concept is that the modules can be assembled and tested separately (by different contractors in different countries if required) without heed to the constraints of the other modules. Final **integration** and testing follow at a later stage.

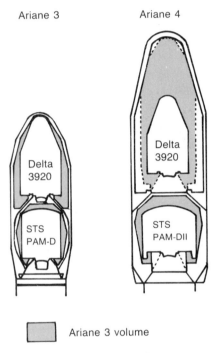

Ariane 3   Ariane 4

Delta 3920

Delta 3920

STS PAM-D

STS PAM-DII

Ariane 3 volume

Figure P5 The **payload envelope**s of the **Ariane** 3 and 4 **launch vehicle**s compared with those for the **Delta** and **Space Shuttle/PAM**. [Arianespace]

PAYLOAD SHROUD

See **shroud**.

PAYLOAD SPECIALIST

A category of person who is part of the crew of the American **Space Shuttle**, but has not been trained as an **astronaut**; an engineer or scientist responsible for a particular **payload** carried by the Shuttle **orbiter**.
[See also **mission specialist**.]

PCM

See **pulse code modulation**.

PEAK

The beam centre or RF **boresight** of an **antenna radiation pattern**; the point on an **antenna gain** plot (a graphical representation of antenna gain) at which the gain is a maximum [see figure B2]. A distinction is made between the **radio-frequency** boresight referred to here and the geometrical boresight (axis) of the antenna, the angle between the two being known as the **squint** angle.

Also used as a verb (such as in 'to peak', 'to peak up', etc), particularly when aligning an **earth station** antenna with a satellite in **orbit**. [See also **beamwidth**.]

PERCUSSION PRIMER

A mechanical device in an **ignition system**, whereby the impact of a firing pin against an anvil ignites an impact-sensitive charge, which then ignites the **propellant**.

PERFORMANCE

In general, a term used to express the quality of functioning of a component or **system** (e.g. '**payload** performance', 'RF performance'). The performance of a **rocket engine** or **rocket motor** defines that of the **stage** or total **launch vehicle** to which it is attached: it is typically measured by quantities such as **thrust** and **specific impulse**. [See also **uprate**.]

PERIAPSIS

The point in an **orbit** closest to the centre of gravitational attraction; the opposite of **apoapsis**. See **perigee**, **perihelion**.

PERIGEE

The point at which a body orbiting the Earth, in a **elliptical orbit**, is at its closest to the Earth [figure G1]; the opposite of **apogee**. The word is derived from the Greek 'perigeion': the prefix 'peri' meaning 'near or adjacent'; the suffix 'gee' referring to 'Earth'. [See also **orbit**, **perigee kick motor (PKM)**.]

PERIGEE KICK MOTOR (PKM)

A **rocket motor** used to place a satellite in a **geostationary transfer orbit** (GTO), fired when the spacecraft reaches the point in the GTO ellipse closest to the Earth, the **perigee**. Sometimes called a perigee stage.

Geostationary satellites launched by the American **Space Shuttle** require a PKM, since the Shuttle delivers spacecraft to a **low Earth orbit**

(most spacecraft launched by the Shuttle require some form of **upper stage**) [see figures P2 and P3]. Some **expendable launch vehicle**s (ELVs) also require the use of a PKM (e.g. **Delta**); others (e.g. **Ariane**) inject their **payload**s directly into GTO using their uppermost **stage**, thus removing the need for a PKM.
[See also **payload assist module (PAM), inertial upper stage (IUS), Centaur G, apogee kick motor (AKM), rocket engine.**]

PERIGEE STAGE

A **launch vehicle** stage which injects a spacecraft into a **transfer orbit** by firing its **engine**(s) at the **perigee** of that **orbit**. Sometimes used as an alternative term for **perigee kick motor**.
[See also **stage.**]

PERIHELION

The closest point to the Sun in an elliptical solar orbit (from the Greek for 'near to the sun'); the opposite of **aphelion**. See **perigee**.

PERSONAL RESCUE SPHERE (PRS)

A self-contained, one-man **life support system** which allows crew members not protected by pressure suits to be transferred from a disabled Space Shuttle **orbiter** to a rescue vehicle. The PRS is an inflatable sphere 0.86 m in diameter.

PERTURBATIONS (of a SPACECRAFT ORBIT or TRAJECTORY)

Disturbances which cause a deviation in a **spacecraft**'s nominal orbital path or **trajectory**.
   There are three main perturbation mechanisms that are particularly important for a satellite in **geostationary orbit** [see **triaxiality, luni–solar gravity** and **solar wind**]. More important for a satellite in **low Earth orbit** is **atmospheric drag**.

PFD

See **power flux density**.

PGSE

An acronym for payload ground support equipment. See **ground support equipment**.

PHASE

The position in the cycle of a periodic quantity expressed as an angle (e.g. 'the two waveforms are 90 degrees out of phase'). See, for example, **phase modulation**.

PHASE MODULATION (PM)

A transmission method using a modulated **carrier** wave, whereby the phase of the carrier is varied in accordance with the amplitude of the input signal. The PM technique relies on the fact that two or more samples of the same periodic waveform (or carrier) can be arranged to have different time origins (i.e. they are 'out of phase'). In the **demodulation** process the phase changes can be detected and the original signal reconstructed. For example, carrier waveforms 180 degrees out of phase can be used to represent the digits '0' and '1', a technique commonly referred to as PSK modulation, phase shift keying, or binary phase shift keying (BPSK).
[See also **modulation, amplitude modulation (AM), frequency modulation (FM), QPSK.**]

PHASE SHIFT KEYING (PSK)

See **phase modulation (PM)**, and **QPSK.**

PHOTON

A quantum of **electromagnetic radiation**; an indivisible unit of electromagnetic energy. The energy of a photon is the product of the **frequency** of the radiation and Planck's constant. Thus, for example, the energy of an x-ray photon is greater than that of an infrared photon.
[See also **electromagnetic spectrum.**]

PHOTON ENGINE

A hypothetical **rocket engine** in which **thrust** is created by a flow of **photon**s. Such a device has the highest **specific impulse** imaginable since, by definition, the photon travels at the velocity of **light**.

PHOTOVOLTAIC CELL

See **solar cell**.

PIN-WHEEL LOUVRE(S)

A device used in the thermal control of a spacecraft to shield or expose a **radiator** surface; an early example of a **louvre** system comprising a circular, segmented wheel controlled by a bimetallic spring which rotates to enhance or inhibit radiation of thermal energy.

PIONEER

A series of American planetary exploration **probe**s launched towards

Venus, Jupiter and Saturn. Earlier spacecraft bearing the same name were placed in Earth orbit: one of these, Pioneer 3, led to the discovery of the **Van Allen belts**.

PITCH

A rotation about the pitch axis—see **spacecraft axes**.
[See also **pitch-over**.]

PITCH AXIS

See **spacecraft axes**.

PITCH-OVER

The moment at which a **launch vehicle** leaves the vertical part of its **launch profile** and begins a slow rotation about the **pitch axis**, which decreases the angle between its longitudinal axis and the ground and allows it to follow a **trajectory** which will culminate in an orbital **injection**.

PIXEL

An abbreviation for 'picture element', usually referred to an electronically generated **image**, where a pixel is the smallest definable element of picture information. For instance, if an image scanning system has a **resolution** of 400 horizontal lines and each line contains 500 individual **bit**s of picture data, the total number of pixels will be 200 000. Thus the number of pixels gives an indication of the definition, or clarity, of the image. Pixel size is analogous to the grain size of photographic emulsion, except that the latter is neither constant across the image nor regularly organised in the manner of a grid.

PKM

See **perigee kick motor (PKM)**.

PLANE CHANGE

A change in the plane of an **orbit**; a change in the **orbital inclination**.
These **manoeuvre**s, usually made using **reaction control thruster**s, are kept to a minimum, because they involve a relatively high **propellant** consumption. Since the plane into which a spacecraft is initially launched is related to the latitude of the launch site, it is desirable for the two to be as closely matched (co-planar) as possible. Many satellites are required to operate from **geostationary orbit**: for injection into this and other **equatorial orbit**s, a launch site on the equator would be ideal, since no plane change would be necessary. For

this reason, the majority of the West's launch sites are located as close to the equator as possible: e.g. **Kennedy Space Center** at 28.3° north in Florida, the USA's most southerly mainland state, and the **Guiana Space Centre** near Kourou in French Guiana at 5.23°N. A launch from the latter offers a more efficient route to GEO due to the smaller plane change. It also makes greater use of the Earth's inherent (west to east) rotational velocity, which is greatest at the equator (about 1670 km s⁻¹); the corresponding velocity at latitude L is 1670 cos L.

Since most of the USSR's territories are far from the equator, early Soviet **communications satellite**s were launched into the highly elliptical **Molniya orbit** rather than GEO.

PLANETARY ALBEDO

See **albedo**.

PLANETARY BODY

Any astronomical body in **orbit** around a star, including solid and gaseous planets, moons and asteroids; a subset of **celestial body**.

PLASMA

An ionised gas, i.e. a mixture of positively charged ions and negatively charged **electron**s. See **ionisation**.

PLASMA ROCKET

See **electric propulsion**.

PLASMA SHEATH

A layer of ionised air surrounding a spacecraft re-entering the Earth's atmosphere, generated by frictional heating. It effectively blocks all radio communication with the spacecraft.
[See also **S-band blackout, ionisation, ionosphere**.]

PLATFORM

A general term for a spacecraft, or part of a spacecraft, usually with the implication that a stable reference or support structure is being provided (e.g. 'space platform', 'orbital platform', 'observation platform'). More specifically, it can refer to the **antenna module** or **antenna platform** of, for example, a **communications satellite**.
[See also **'bus', inertial platform, launch platform, San Marco platform**.]

PLESETSK

The location of the Soviet **launch site** known as the **Northern Cosmodrome**.

PLSS

An acronym for **portable life support system** and primary life support system or subsystem.

PLUG NOZZLE

An alternative to the bell-shaped **exit cone** used in the **nozzle**s of contemporary **rocket engine**s and **rocket motor**s, whereby a ring-shaped **combustion chamber** discharges combustion gases against the outer surface of a central truncated cone (giving the appearance of a conventional nozzle with a plug in it) [see figure P6]. The ring shape of the combustion chamber and its nozzle outlets gives rise to the alternative term 'annular nozzle'.

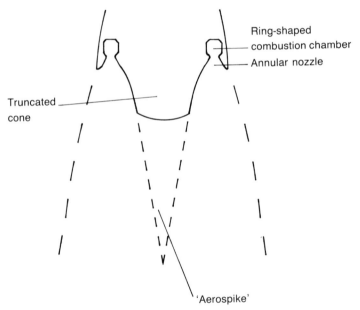

Figure P6 Cross section through a **plug nozzle**—an alternative to the bell-shaped **exit cone**.

The concept was developed in the 1960s for possible applications on future **single-stage-to-orbit** (SSTO) launch vehicles, largely because it offered an automatic adjustment to the variation in atmospheric pressure between ground level and the upper atmosphere. For example, if a bell nozzle is designed for optimum efficiency at high altitude (i.e.

in vacuum or near-vacuum conditions), ambient air pressure at lower altitudes will compress the exhaust **plume** and reduce the efficiency of the device [see **flow separation**]. In contrast, the flow from the plug nozzle adapts automatically to the changing external pressure and always operates at or near optimum efficiency. In most designs the central cone was truncated to optimise the vehicle for the heating effects of **re-entry**. The central stream of exhaust gases followed the line of the cone forming a system of shock waves below the nozzle, known as an 'aerospike' (from 'aerodynamic spike').
[See also **expansion ratio**.]

PLUGS-OUT TEST

Jargon: a test of a **launch vehicle** conducted at the **launch pad** with the **umbilical**s disconnected as they would be just prior to a real **launch**.

PLUMBING

Engineering slang for any assembly of **waveguide**-based components. The term derives from the resemblance of lengths of waveguide, bolted together at their flanges, to industrial pipework. Also used for the piping in a liquid **propulsion system**.

PLUME

A stream of **combustion** products ejected from a rocket exhaust **nozzle**. [See also **plume impingement**, **degradation**.]

PLUME IMPINGEMENT

An undesired contact between a **thruster** plume and a spacecraft component or surface.

The main effect is the production of unwanted forces which disturb the **attitude control** of the spacecraft: the impingement of a thruster plume on the **solar array** panels of a **three-axis-stabilised** spacecraft produces a so-called 'windmill torque'. The effect can be reduced by careful siting of the thrusters on the spacecraft body, but this is difficult since the exhaust gases are not emitted in a narrow, pencil beam (about 90% of the gas flow is released in a 60° cone). As a solution, thruster firings can be timed for favourable array positions, or a system for automatic correction of the disturbances can be used.

In addition to unwanted torques, thruster plumes can produce heating effects, erosion due to high-speed droplets of unburnt **propellant**, corrosion due to the propellant and its combustion products, and the contamination of optical surfaces.
[See also **plume shield**, **degradation**.]

PLUME SHIELD

A device mounted adjacent to a spacecraft's **rocket engine, rocket motor** or **thruster**, designed to protect the spacecraft surfaces from the heating effects of a rocket's **plume** and the **combustion** products, which would otherwise cause **degradation** of its **surface coatings**.
[See also **plume impingement**.]

PM

See **phase modulation**.

PMD

See **propellant management device**.

POGO

A rhythmic longitudinal (vertical) oscillatory motion (or vibration) experienced by a **launch vehicle**. It can be generated by an interaction between the **propellant** flow and **combustion** process, and the structure of the vehicle. The term is derived from a similarity with the motion of a pogo stick.
[See also **sloshing**.]

POGO SUPPRESSOR

A device which absorbs the longitudinal vibrations that can occur during the **flight** of a **launch vehicle** [see **pogo**]. Typically a sealed tube containing a liquid which damps out the oscillation through friction with its container (in a similar way to the **nutation damper** used on some spacecraft).

POINTING

(i) The process of aligning a communications **antenna** or an astronomical telescope ('an antenna requires pointing...').

(ii) An attribute of an antenna ('the antenna pointing is good, so optimum performance is assured...').

[See also **pointing loss**.]

POINTING LOSS

A loss in signal power in a communications link due to the mispointing of **antenna**s. The concept is applicable to all links: terrestrial, Earth-to-space and space-to-space. The example below concerns the **uplink** and **downlink** paths of a **satellite communications** system.

The **earth station** antenna cannot always remain pointed precisely at the **satellite** because the spacecraft does not remain stationary with respect to the earth station (see **perturbations** and **box**). This means that the satellite is not located on the **boresight** of the ground-based antenna and does not receive the peak power transmitted from the earth station. The degree of loss depends on the deviation from boresight. In addition the satellite antenna is not always maintained accurately pointing towards the earth station because it requires the expenditure of **reaction control thruster** fuel (but see **RF sensor**). This reduces the received power even further.

By the same token, since the earth station is not pointed directly at the satellite and is not on the boresight of the spacecraft antenna, it too receives a lower signal power.

[See also **beamwidth**.]

POLAR MOUNT

A structure for the support and guidance of an **antenna** which has its axis aligned with that of the Earth, with the result that steering along the **geostationary arc** is possible by rotation about this single axis (as opposed to the two-axis steering required with the alternative **altitude-azimuth mount**). Another name for a mount which has its lower axis parallel to the Earth's axis is hour angle — declination or 'HA–DEC'. When used for astronomical telescopes it is called an **equatorial mount**.

POLAR ORBIT

An **orbit** at a high **inclination** to the equatorial plane of a **planetary body**, such that a **satellite** in that orbit passes over the body's polar regions [see figure H2]. The **orbital track** of the satellite does not have to cross the poles exactly for the orbit to be defined as 'polar': orbital tracks which pass within 20 or even 30 degrees of the pole are still classed as polar orbits.

POLARISATION

The physical phenomenon whereby electromagnetic waves are restricted to certain directions of vibration. There are two main types of polarisation: linear, in which the electric vector of the radio wave is confined either to the vertical or the horizontal plane; and circular, where the electric vector is rotated either clockwise or anticlockwise to give right- or left-hand circular polarisation (RHCP or LHCP).

The orientation of the vertical and horizontal polarisation vectors depends on the particular communications system: for instance, 'horizontal' may be defined as parallel to or at 45 degrees to the equatorial plane. The precise definition of circular polarisation depends on its source: the CCIR recommendation (used in Europe) defines polarisation 'looking away from the transmitter' whereas the IEEE definition, used in the United States, defines it 'looking towards the transmitter'.

With circular polarisation the **antenna feed**, from which the RF radiation originates, can be designed to cause the electric field vector to rotate clockwise or anticlockwise. Only a small amount of the radiation of one polarisation is receivable in a feed designed for the opposite hand, so polarisation represents an effective method of **interference** reduction. (Although opposite polarisations can give as much as 30 dB of **isolation** between channels on the same frequency, this can be degraded, away from the axis of the beam centre, by conversion to elliptical polarisation.) See **polarisation conversion**.

In satellite communications, signals can be transmitted on opposite polarisations in order to double the use of a **frequency band**. This is known as **frequency reuse**.

POLARISATION CONVERSION

A reversal of **polarity** suffered by an **RF wave** on its passage through the atmosphere under adverse weather conditions or within a **feed** or **antenna** system.
[See also **cross-polar(ised)**.]

POLARISATION FREQUENCY REUSE

See **frequency reuse**.

POLARISER

A device attached to the **feedhorn** of an **antenna** which polarises the RF radiation. See **polarisation**.

POLARITY

In **radio-frequency** communications, an atttribute of an **RF wave** whereby it may be assigned one of two opposite polarisations to provide **discrimination** between similar signals which might otherwise interfere. Used in the sense of 'RF waves of opposite polarity...'. See **polarisation**.

POLYBUTADIENE

A type of solid **fuel** used in **rocket motor**s, with a chemical formula of the type: $CH_2CHCHCH_2$. Two variants used in **solid rocket booster**s are carboxyl terminated polybutadiene (CTPB) and hydroxyl terminated polybutadiene (HTPB) (also written as carboxy-terminated and hydroxy-terminated polybutadiene). CTPB has the monovalent group —COOH added to the basic formula and HTPB contains the —OH group.

[See also **solid propellant**.]

PORTABLE LIFE SUPPORT SYSTEM (PLSS)

A **spacesuit** 'back-pack' which contains everything necessary to maintain a comfortable working environment within the suit (i.e. an oxygen supply for breathing and suit pressurisation, and a water supply for thermal control and drinking). The suits used on the American **Space Shuttle**, for example, contain supplies for up to seven hours in the suit, which includes a nominal six-hour EVA, a 30-minute reserve and a secondary 30-minute supply for use as a back-up in an emergency. Astronauts can also 'plug themselves into' the **orbiter** supplies in the **airlock**.

PLSS is commonly pronounced 'pliss' and is also an acronym for primary life support system or subsystem: the main supply in the portable life support system is known as the primary life support subsystem.

[See also **pre-breathe**.]

POWER

(i) A measure of the 'rate of doing work', measured in watts (W); or the energy per unit time ($J\,s^{-1}$).

(ii) More colloquially, in both engineering and everyday use, a 'commodity' that can be generated, stored and used—more properly termed '**energy**' (e.g. as in 'nuclear power', 'wave power', 'power supply', etc).

[See also **AC power**, **DC power**, **RF power**.]

POWER-AUGMENTED HYDRAZINE THRUSTER (PAHT)

See **hiphet thruster**.

POWER BUDGET

A method of accounting for the **power** requirements of a spacecraft or subsystem with respect to the capability of the **power supply** to provide that power; a sum of subsystem power requirements.

During a spacecraft project, each subsystem engineering discipline aims to design its hardware within its allotted portion of the power budget. The power budget allows changes in the overall spacecraft power requirement to be monitored throughout the project.

[See also **mass budget**, **propellant budget**, **link budget**.]

POWER BUS

The main electrical power distribution circuit in a spacecraft. Usually abbreviated to '**bus**', a short form of 'bus-bar'.
[See also **power, bus voltage.**]

POWER FLUX DENSITY (PFD)

The level of **radiated power** (measured in watts) received per square metre (units: dBW m^{-2}).

Although relevant for any radiated-telecommunications system (between any two points), PFD is most often quoted from the viewpoint of a receiving **ground station**, which has a fixed effective **antenna** aperture (measured in m^2). If the PFD provided by a spacecraft is too low, only a larger antenna will be able to collect sufficient power to constitute a 'good signal'.

PFD may be calculated using the expression

$$PFD = 10\log_{10} \frac{PG}{4\pi R^2}$$

where P is the **transmitter** power (W), G is the **antenna gain** (expressed as a pure number) and R is the distance (m) between the spacecraft and the Earth. Removing the gain, G, from the expression would give the PFD for an **isotropic antenna**. The product PG is the equivalent isotropic radiated power (EIRP).

If P and G are available in decibels, the PFD would be $(P + G - 162.1)$ for a spacecraft in **geostationary orbit** where R is about 3.6×10^7 m.
[See also **equivalent isotropic radiated power (EIRP), decibel.**]

POWER-LIMITED

Jargon: 'limited in terms of available **power**'. The power consumption of all components and **subsystem**s in any **spacecraft** design is constrained by the capability of its **power supply** and must be tailored to suit [see **power budget**]. By way of illustration, a **communications satellite** designed to carry a given number of **transponders**, but with insufficient power to operate them simultaneously, would be 'power-limited'.
[See also **mass-limited.**]

POWER SUPPLY

Any device which provides electrical power, usually matched to the specific requirements of the device or system in receipt of that power. See **power**.
[See also **primary power supply, secondary power supply, regulated power supply, unregulated power supply.**]

PRE-BREATHE

The practice of breathing pure oxygen in preparation for an EVA in a **spacesuit** with a low-pressure, pure oxygen supply. For example, the suits used on the American **Space Shuttle** are pressurised to only 27.6 kPa, 4 psi (cf 101.4 kPa, 14.7 psi, Earth's surface and nominal **orbiter** pressure). This would normally require pre-breathing of 3.5 hours prior to EVA to eliminate **decompression sickness** ('the bends') when exposed to the lower pressure. This time has, however, been reduced in most cases to about 40 minutes by reducing the orbiter's **cabin pressure** to about 70 kPa (10 psi) 12 hours before an EVA. A 'zero pre-breathe suit' has been proposed for use with the international **Space Station**: it would use oxygen at a pressure closer to cabin pressure and would have to be stronger as a result.

PRECESSION

The motion of a spinning body (e.g. a planet, **satellite** or **gyroscope**) whereby its axis of rotation sweeps out a cone (like a wobbling spinning top) [see figure N3]; also called 'spin-axis precession'.
[See also **nutation**.]

PRE-EMPHASIS

A method of improving the **signal-to-noise ratio** in a **frequency modulation** system by increasing the deviation at higher **baseband** frequencies.

PRE-OPERATIONAL

Generally: 'prior to becoming capable of or actually involved in operations'.

For example, with reference to satellite communications systems this means that there are insufficient resources (i.e. satellites and satellite **channel**s) to offer a continuous and guaranteed service. See **operational**.
[See also **in-orbit spare**, **ground spare**.]

PRE-PREG

An abbreviation for 'pre-impregnated': generally refers to a woven fibre material, used in the manufacture of low-density spacecraft structures, which has been pre-impregnated with a matrix material [see **carbon composite**].

In a process known as 'laying up', the pre-preg is applied in layers to a former, which governs the final shape. The 'lay-up' is then cured in an autoclave, under conditions of high temperature and pressure. This

process, whereby a material hardens permanently after an application of heat and pressure, is known as thermosetting.
[See also **tape wrapping**.]

PRESSURANT

A pressurising gas; a gas (typically helium) which provides the pressure (in a pressure-fed propulsion system) to force **propellant** from its storage tank to a **rocket engine**. In some **launch vehicle** stages, where the outer **skin** of the **stage** itself forms the **propellant tank**, the pressurant also helps to keep the tank rigid.
[See also **liquid propellant, ullage pressure, blowdown system**.]

PRESSURE EXPANSION RATIO

See **expansion ratio**.

PRESSURE SUIT

See **spacesuit**.

PRESSURISATION

(i) In pressurised **propulsion system**s, the act of increasing the pressure of the vapour above a **liquid propellant**, or of introducing another **pressurant**, chiefly to provide a force to deliver the **propellant** to an outlet valve (particularly the sequence of operations which pressurises the **propellant tank**s of a **launch vehicle** prior to **lift-off**).
[See also **ullage vapour**.]

(ii) The pressure to which a **spacesuit** or the cabin of a **manned spacecraft** is 'pressurised'.
[See also **decompression, pre-breathe**.]

PRIMARY LIFE SUPPORT SYSTEM

See **portable life support system (PLSS)**.

PRIMARY POWER SUPPLY

The main source of **DC power** available aboard a spacecraft. For most Earth-orbiting satellites, it is the **solar array**, used whenever the spacecraft is in sunlight. Some manned space vehicles use the **fuel cell** as the primary source, while interplanetary spacecraft venturing beyond the orbit of Mars use the **radioisotope thermoelectric generator** (RTG).
[See also **secondary power supply**.]

PRIME FOCUS

The point of focus of the primary **reflector** of a communications **antenna** or an astronomical telescope. The position for the **feedhorn**, recording device (film or detector), or secondary reflector, depending on the complexity of the reflector system. See **Cassegrain reflector**, **Gregorian reflector**, **Newtonian telescope**.

PRIME SPACECRAFT

The **spacecraft** intended for use on a particular **mission** (as opposed to a **back-up** spacecraft).
[See also **ground spare**, **in-orbit spare**.]

PRIME SYSTEM

Any **system** designated as the main provider of a particular function or operation (as opposed to a **back-up** or **redundant** system).

PROBE

(i) A colloquial term for an **unmanned spacecraft**, usually designed for scientific investigation of another planetary body (e.g. the **Viking** and **Voyager** planetary probes) or other celestial object (e.g. **Giotto**). Scientific, and other, spacecraft in Earth orbit tend to be termed **satellite**s.

(ii) To examine with a probe; any device designed to 'probe' or test a medium under investigation (e.g. the surface of a planetary body, a planetary atmosphere, an electromagnetic field, etc).

(iii) The male part of a docking mechanism. See **docking probe**.

PROCESSING

A blanket term for the preparation of a **spacecraft** or **launch vehicle** for **spaceflight**, used particularly with regard to the American **space transportation system** (e.g. 'Shuttle processing', '**orbiter** processing'). See **orbiter processing facility** (OPF).

PROGRADE ORBIT

An **orbit** in which the orbiting body has the same sense of rotation as the orbited body (also known as a 'direct orbit'); the opposite of **retrograde orbit**.

PROPELLANT

Any substance or combination of substances which constitute a mass to be expelled at high velocity to produce a propulsive reaction force or **thrust**.

Propellants comprising a single component are called monopropellants. Propellants with two components (a **fuel** and an **oxidiser**) are called **bipropellant**s.

Propellants can also be classified according to their physical state, either solid, liquid or gaseous. **Solid propellant**s generally consist of a mixture of fuel and oxidiser (they are considered as bipropellants by some because they have two components, but as monopropellants by others since they are handled and used as a single-component propellant). **Liquid propellant**s may be monopropellants or bipropellants (the latter are stored separately and combined in a **combustion chamber**). Although solid and liquid propellants are usually employed separately, they are used together in the **hybrid rocket**, the most common combination being 'solid fuel + liquid oxidiser'. The monopropellants used in the **cold-gas thruster** are unusual in that they are generally stored as a gas.

Propellants are also classed as either 'storable' or 'non-storable', the distinction being one of temperature. If a propellant retains its physical properties at normal terrestrial environmental temperatures, it is 'storable'; if it needs to be cooled significantly below these temperatures to render it usable in a propulsion system, it is 'non-storable'. All solids and some liquids are storable; **cryogenic propellant**s are non-storable.

[See also **hypergolic, electric propulsion, nuclear propulsion, rocket engine, rocket motor**.]

PROPELLANT BUDGET

A method of accounting for the amount of **propellant** in a spacecraft's **propulsion subsystem** available for the various phases of its operational **lifetime**. More colloquially termed the 'fuel budget'.

Propellant is required, by various different types of **spacecraft**, for orbital **injection, plane change**s, **mid-course correction**s, **attitude control, orbital control, orbital relocation** and **de-orbit**ing. About half of the launch mass of a typical **satellite** destined for **geostationary orbit** consists of propellant, the majority of which is used for the **apogee** burn.

[See also **fuel, oxidiser, liquid propellant, solid propellant, apogee kick motor, reaction control thruster, retro-rocket**.]

PROPELLANT CHARGE

See **propellant grain**.

PROPELLANT DISPERSAL SYSTEM

An arrangement of **pyrotechnic** charges mounted on a **launch vehicle** stage which can be used in an emergency to open the **propellant** tank(s), disperse the propellant and destroy the vehicle. Charges are typically arranged in a continuous line and tend to 'un-zip' the tank when ignited.

PROPELLANT GRAIN

A quantity of **solid propellant** shaped to give the required combustion characteristics (namely, thrust against time) for a particular **rocket motor**. Also called a propellant 'charge'.

The **thrust** available from a block of propellant is proportional to the area of the combustion surface. One method of burning the propellant is by 'cigarette combustion', where the propellant block is ignited at one end and burns to the other like a cigarette (also known as 'end-burning'). In this case the cross section of the block gives the surface area of active combustion and limits the thrust. The active surface area can be increased by a cylindrical hole through the centre of the block so that, as the propellant burns, the hole enlarges radially and the thrust increases. Constant thrust, which is more often required, can be provided by a central hole which is star-shaped in cross section [see figure P7]. A variation in thrust throughout the **burn** can be arranged by varying the cross section along the length of the propellant grain. This is generally a requirement for a **launch vehicle** solid rocket motor (SRM) which needs to reduce thrust during the period of **maximum dynamic pressure**.

[See also **propellant, ignition system**.]

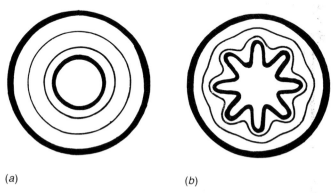

(a)   (b)

Figure P7 Solid **propellant grain** cross sections: (a) thrust increases throughout burn, (b) constant thrust.

PROPELLANT LINER

A material placed between the **motor case** and the **propellant** in a solid rocket motor which protects exposed parts of the case from the heat of combustion and supports the **propellant grain** during manufacture, as it is poured into the case. Also known as a '**flame inhibitor**' [see figure R5].
[See also **rocket motor**.]

PROPELLANT MANAGEMENT DEVICE (PMD)

A device or structure inside a **propellant tank** which uses surface tension (capillary) forces to position and hold the **propellant** over the tank outlet, thus allowing propulsion in weightless conditions (i.e. in the absence of inertial or gravitational forces). PMDs vary in design from simple vanes and perforated metal sheets to more complex tubular structures with circular openings, known as collectors or galleries.
[See also **ullage rocket**.]

PROPELLANT MASS FRACTION

See **mass ratio**.

PROPELLANT TANK

Any container or reservoir in a **spacecraft** or **launch vehicle** (or on the ground as part of the **ground support equipment**) used to hold **liquid propellant** (**fuel** or **oxidiser**). The equivalent item for **solid propellant** is termed a **motor case**.
[See also **external tank, vent and relief valve, pressurant, ullage, pogo, sloshing**.]

PROPULSION BAY

The part of a **launch vehicle** or **spacecraft** which houses the **rocket engine**s or **rocket motor**s, typically incorporating a **thrust frame** and bordered by engine **cowling**s and other **fairing** components. Also called an engine bay.

PROPULSION MODULE

A self-contained section of a modular spacecraft containing the propulsion subsystem [see figure M2].

For a **communications satellite** the propulsion module usually comprises an **apogee kick motor** (AKM), and the **propellant** and hardware for an **attitude and orbital control system** (AOCS), but see

combined propulsion system.
[See also **service module, payload module.**]

PROPULSION SUBSYSTEM

A spacecraft **subsystem** which provides the spacecraft's propulsion requirements throughout its **lifetime**. It may use either **solid propellant, liquid propellant** or gaseous propellant (or a combination as in the **hybrid rocket**).
[See also **propulsion system, rocket motor, rocket engine, ignition system, combustion chamber, nozzle, reaction control thruster, cold-gas thruster, combined propulsion system, electric propulsion.**]

PROPULSION SYSTEM

Any **spacecraft** or **launch vehicle** system that provides propulsion. Compared with **propulsion subsystem**, 'propulsion system' is the more general term: a launch vehicle 'en masse' can be considered a 'propulsion system'; a 'propulsion subsystem' is a part of a vehicle (as is its guidance subsystem, **thermal control subsystem**, etc). See **subsystem**.
[See also **chemical propulsion, electric propulsion, electrothermal propulsion, nuclear propulsion, rocket engine, rocket motor, propellant, combined (bipropellant) propulsion system.**]

PROTECTION RATIO

In **satellite communications** systems, the ratio of the power of the wanted signal to that of the unwanted (interfering) signal (C/I, for **carrier**-to-**interference** ratio). It is measured in decibels (dB); e.g. if the value of the protection ratio is 3 dB, the wanted signal has twice the power of the unwanted signal. This is for 'single-entry' interference (only one interferer); for more than one interfering signal, the ratio is simply the power of the wanted signal divided by the 'power sum' of the interfering powers.

PROTON

A Russian **launch vehicle** developed in the 1960s and first flown in 1968. Also known as the SL-12. Different variants have two, three or four **stages**: the three-stage version, offered to the West as a commercial launcher in the late 1980s, has a **payload** capability of 1500 kg to **geostationary orbit** (GEO).
[See also **Energia.**]

PROVING STAND

A ground-based facility for testing (and 'proving the design' of) **rocket** engines and **rocket motor**s. Also called a test stand [see frontpiece].

PRS

See **personal rescue sphere**.

PSK

An acronym for phase shift keying. See **phase modulation**.

PTT

An acronym for Posts, Telegraph and Telecommunications organisation; a country's provider of **telecommunications** and other associated services.

PULSE CODE MODULATION (PCM)

A transmission method using a modulated **carrier** wave, whereby the amplitude of the original analogue input signal is sampled at discrete time intervals to create a representative digital translation of the signal. PCM was devised specifically to enable (analogue) speech to be transmitted in a digital form, using the principle that, if an analogue signal is sampled at a rate at least twice that of the highest frequency present, the original speech signal can be reconstructed with acceptable quality from the discrete sample values.

Most PCM systems conform to **CCITT** recommendations and sample at a rate of 8 kHz (8000 times a second). Each sample provides a voltage level which is encoded into a seven-**bit** code, allowing 32 possible values, an eighth bit being added to represent the sign. The **bit rate** for a PCM channel is therefore $8 \times 8000 = 64$ kbits s^{-1}.

PULSED THRUST

A **thrust** produced in the operation of a **rocket engine**, specifically a **reaction control thruster**, which consists of a sequence of short bursts; an operational alternative to a continuous thrust. Pulses as short as about 12.5 ms are possible, but the minimum duration is limited by the time taken for the thruster valves to open and close (about 5 ms in each case), since this limits the repeatability of the process.

PUMPED FLUID LOOP

See **cooling system**.

PURGE

The removal of contaminants (e.g. 'to purge a **propellant tank**'). Dry nitrogen and helium are gases commonly used for purging a **launch vehicle**'s tanks prior to **propellant** loading and **lift-off**. The **payload bay** and parts of the fuselage of the American Space Shuttle **orbiter** are purged both 'pre-flight' and 'post-flight' to remove contaminants and toxic gases, and to maintain specified temperatures and humidities. The purge gas is monitored for contaminants by a mass spectrometer.

PYROPHORIC FUEL

A **fuel** that ignites spontaneously in air (compare with **hypergolic**).

PYROTECHNIC CABLE-CUTTER

A type of pyrotechnic **actuator**; a device which uses an explosive charge to sever a cable connecting two items of **hardware** (e.g. for releasing **solar array** panels folded against the side of a **satellite**). A pyrotechnic bolt-cutter performs the same function for different applications (e.g. separating the **stages** of a **launch vehicle**).

PYROTECHNIC VALVE

A type of pyrotechnic **actuator**; a device which uses an explosive charge to either close or open a liquid or gas supply-line (e.g. to isolate a **liquid apogee engine** from the rest of a **combined propulsion system** after the **apogee** firing(s) have been made).

PYROTECHNICS

A class of devices which use an explosive charge to:

(i) separate one item of **hardware** from another [see **pyrotechnic cable-cutter**], or isolate part of a **propulsion system** [see **pyrotechnic valve**]; or

(ii) ignite the **propellant** in a **rocket motor** or **rocket engine** [see **ignition system**].

Q

QM

An acronym for qualification model (often abbreviated to 'qual model'); a version of a spacecraft constructed to verify its **performance** to a given set of specifications, to which it must comply to be deemed **space-qualified**.

QPSK (QUADRATURE PHASE SHIFT KEYING)

A method for modulating the **radio-frequency** carrier in digital **satellite communications** links; a type of **phase modulation**. Also called *quaternary* phase shift keying.

In two-phase **modulation**, where sample **carrier** waveforms are 180 degrees out of phase, one **bit** of data is encoded on the carrier for each phase change, allowing the presentation of a '1' or '0'. This is alternatively known as binary phase shift keying (BPSK). By employing more than two phases it is possible to encode more than one bit per phase, thereby increasing the **bit rate** without altering the modulation rate. Thus four-phase modulation (QPSK) codes two bits per phase change allowing the representation of '00', '01', '11' and '10' for phase changes of 0,90,180 and 270 degrees respectively. This allows **MODEM**s employing QPSK to generate and transmit bit streams at burst rates of 66 Mbits s^{-1} or greater. [See also **burst, modulation**.]

QUALIFICATION

See **space-qualified**.

QUASI-DBS

See **direct broadcasting by satellite**.

QUATERNARY PHASE SHIFT KEYING

See **QPSK**.

R

RADAR

An acronym for radio detection and ranging; a method for the determination of an object's distance and direction using **radio-frequency** radiation. Distance is derived by measuring the time the reflected energy takes to return to the transmitter. Radar can be used from space for Earth **remote sensing** (e.g. ocean wave heights) and planetary surface mapping (e.g. determination of the topography of Venus despite the visual opacity of its atmosphere).

RADAR ALTIMETER

An instrument included on some **remote sensing** and planetary exploration spacecraft which uses radar techniques to ascertain the form of the planet's surface features (topography) by measuring their relative height.

RADIAL THRUSTER

A **reaction control thruster** on a **spin-stabilised** spacecraft mounted in alignment with a radius of the drum (colloquially and collectively termed 'radials'). It produces **thrust** in a direction normal to the spin axis and is typically used for east—west station keeping and orbital repositioning [see **orbital control**.]
[See also **axial thruster**.]

RADIATED POWER

Radio-frequency power emitted from a radiating **antenna**; **electromagnetic radiation** propagating through **free space**. See **equivalent isotropic radiated power (EIRP)**, **power flux density (PFD)**.

RADIATION EFFECTS

See **single-event effect (SEE)**.

RADIATION PATTERN

See **antenna radiation pattern**.

RADIATOR

A device for the removal of heat from a **spacecraft** or components of a spacecraft, usually in the form of a radiator panel, possibly with a number of fins to increase the local surface area.

Radiator panels are of solid or honeycomb construction and are usually covered with **second-surface mirror**s (SSMs) on the outside [see figure S2]. The heat source (e.g. **TWTA**, output **multiplexer**, **battery**, etc) is attached or coupled to the interior surface, which is generally painted black to encourage heat transfer. On a **three-axis-stabilised** spacecraft the radiator panels are usually mounted on the north and south faces since they receive solar radiation obliquely; **spin-stabilised** spacecraft generally radiate from their north and south ends, but if additional radiator area is required extra panels are located about the midriff of the drum [see figure S18].

[See also **heat rejection, thermal control subsystem, thermal doubler**.]

RADIO (COMMUNICATIONS) BLACKOUT

See **S-band blackout**.

RADIO FREQUENCY (RF)

The **frequency** of a radio **carrier** wave: any frequency which lies in the range 10 kHz to 300 GHz ($1 \times 10^4 - 3 \times 10^{11}$ Hz); frequencies used for radio transmissions, both terrestrially and extraterrestrially. The upper limit is flexible, depending on whether currently available or theoretically possible radio equipment is considered.

To differentiate between their different technologies and measurement techniques, electronics and microwave engineers consider the upper limit of 'RF' to be around 1 GHz, where **microwave** takes over; frequencies below about 10 kHz are referred to as 'DC' (direct current).

[See also **frequency bands**.]

RADIO-FREQUENCY IONISATION THRUSTER

See **ion engine**.

RADIO PROPAGATION

Transmission of **electromagnetic radiation**, particularly at **radio frequency** (RF), through **free space** or along a **transmission line**, etc.

RADIO SPECTRUM

The radio-frequency portion of the **electromagnetic spectrum**. Also called the radio-frequency spectrum.
[See also **radio frequency (RF)**, **frequency bands**.]

RADIOISOTOPE THERMOELECTRIC GENERATOR (RTG)

A spacecraft power source which derives energy from an array of thermocouples heated by the decay of a radioactive isotope.

The RTG is used on spacecraft designed to operate too far from the Sun to use the solar cell as a power generator (e.g. planetary exploration **probes** destined for the outer planets). It is rarely used in the inner Solar System, due to the mass of shielding required and the potential hazard of an uncontrolled **re-entry**. Spacecraft using RTGs are sometimes referred to as 'nuclear-powered'.

The operation of the RTG is based on the **thermoelectric effect**, typically using two different semiconductor materials, one n type and one p type [see **solar cell**]. The 'hot junction' is heated by the radioactive source and the cold junction is attached to a **heat sink**.

RAIN ATTENUATION

The weakening due to rainfall of a radio signal propagating through the Earth's atmosphere. See **atmospheric attenuation**.

RAM

An acronym for random access memory, a computer memory to which the user can add information (as opposed to a read only memory, **ROM**).

RANGE

(i) The distance between a **launch vehicle** or a **spacecraft** and the ground station tracking it.

(ii) The act or process of **ranging**.

(iii) The maximum effective distance of a projectile (e.g. on a **ballistic trajectory**), a vehicle or a radio **signal**.

(iv) A geographical area set aside for **rocket** launches or testing (e.g. the USA's **Eastern Test Range**).

[See also **slant range**.]

RANGE SAFETY SYSTEM

Pyrotechnic devices on a **launch vehicle** used to destroy it in the event of a malfunction which could threaten the safety of the firing **range** and the surrounding area.
[See also **propellant dispersal system**.]

RANGING

The determination of the distance between a **launch vehicle** or a **spacecraft** in **orbit** and the ground station tracking it.
[See also **range**, **tracking**.]

RANKINE (°R)

A temperature scale devised by the Scottish engineer William John Macquorn Rankine (1820–1872), which begins at **absolute zero** but is divided into degrees of equal size to those on the Fahrenheit scale, rather than degrees Celsius as is the **kelvin** scale. The Rankine scale (°R or °Rank) is virtually unheard of in the scientific world, but is sometimes used by engineers.

RARC

An acronym for Regional Administrative Radio Conference, a conference convened under the auspices of the International Telecommunications Union (ITU) to plan **communications** services (particularly with regard to the allocation of radio frequencies). See **WARC**.

RAT'S NEST

Engineering slang for a tangle of electrical wiring.

RAW DATA

Unprocessed data; information or experimental results in their most basic form (e.g. the **telemetry** signals as received from a spacecraft before their conversion into measurements of **attitude** or temperature, which can be readily analysed and understood).

REACTION CONTROL SYSTEM (RCS)

A spacecraft **subsystem** which provides **attitude control** by means of a set of **reaction control thrusters** (typically grouped to provide adjustments in **roll, pitch** and **yaw**).
[See also **orbital manoeuvring system (OMS), Apollo, Gemini, Mercury, X-15**.]

REACTION CONTROL THRUSTER

A small **rocket engine** used to make fine adjustments to the **orbit** or **trajectory** and the **attitude** of a **spacecraft** by the expulsion of gas molecules at high velocity (hence the more colloquial term 'gas jet'). The term 'reaction control' is derived from the principle of 'action and reaction' (Newton's third law): the momentum (mass × velocity) of the gas molecules causes an equal and opposite momentum of the spacecraft.

The shape of the thrusters is similar to that of the larger engines, in that they have a **combustion chamber** with a narrow **throat** leading to a divergent exit cone or '**nozzle**' [see figure R1]. Thruster dimensions are, however, measured in millimetres rather than metres, and the **thrust** produced may be as low as 0.5 N (as opposed to hundreds of kN). There are three main varieties of thruster commonly used in spacecraft control: the **hydrazine thruster**, the **cold-gas thruster** and the **bipropellant thruster**.

[See also **hiphet thruster, attitude control, orbital control, momentum dumping, plume, exhaust velocity, specific impulse, pulsed thrust, orbital manoeuvring system (OMS)**.]

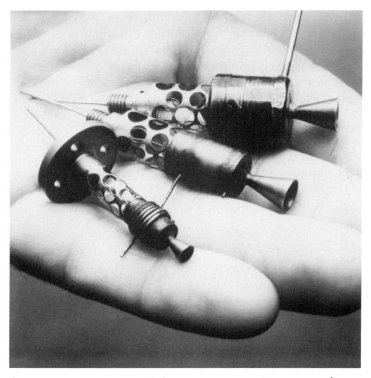

Figure R1 Reaction control thrusters—tiny **rocket engine**s. [British Aerospace]

REACTION WHEEL

A wheel mounted in a **three-axis-stabilised** spacecraft used for **attitude control**.

Reaction wheels are typically fitted in sets of three—one for each of the orthogonal **spacecraft axes**—although it is quite common to install a fourth wheel, mounted at an angle to all three axes, as a spare. If one of the first three wheels fails, the fourth can be used in conjunction with the remaining wheels to provide full three-axis attitude control.

The reaction wheel operates in the following manner. If a power input from its electric motor causes it to spin faster, its angular momentum will increase. This fact alone would infringe the law of conservation of momentum, since the angular momentum of the spacecraft as a whole has increased. However, the law is satisfied by a transfer of angular momentum from the wheel to the spacecraft, which rotates about the wheel-axis in the opposite direction, thereby conserving the total angular momentum of the system. Naturally the spacecraft rotates much more slowly than the wheel, in proportion to their relative moments of inertia. If it was required to rotate the spacecraft in the opposite sense, the wheel would be decelerated: it would lose angular momentum and the spacecraft would gain it by rotating in the opposite direction. In both cases, in the same way that attitude control **thruster** firings require a cancellation force, a further change of wheel momentum would be required to halt the rotation of the spacecraft. It is this 'action and reaction' philosophy that gives rise to the term 'reaction wheel'.

Whereas a **momentum wheel** is normally spinning at a high rate to give stability to the spacecraft, the nominal speed of a reaction wheel is zero and it can be rotated in either direction (typically at up to ± 2500 rpm). A system using reaction wheels is known as a 'zero-momentum' system, as opposed to one using momentum wheels which is called a 'momentum bias' system; both types of wheel are sometimes called inertia wheels.

[See also **momentum dumping**.]

REAL TIME

Jargon: originally the term referred to a data-processing system in which a computer processes **data** as it is generated (i.e. 'in real time'), rather than recording or storing it for later use. The term is now also applied to more general activities.

RECEIVE CHAIN

A collection of electronic and microwave equipment which receives, filters and amplifies a signal fed from a receive antenna. In a spacecraft

communications payload the receive chain generally comprises an **input filter** or **multiplexer**, **low-noise amplifier**, **downconverter**, preamplifier, **channel filter** and associated switches and **hybrid**s [see figure C2]. Whether or not to include the channel filter or the receive antenna in the definition of the receive chain depends on the particular design and, to an extent, personal choice.
[See also **receiver, transmit chain, amplifier chain, input section, output section**.]

RECEIVER

Generally, a device used to detect and amplify incoming electrical signals or modulated radio waves and convert them into signals to drive an output device (e.g. audio or video equipment), or to pass them to further devices prior to retransmission.

In the typical satellite **transponder**, the receiver comprises an **input filter**, which confines the frequency **bandwidth** of the **signal**, several pre-**amplifier** stages, and a **mixer** or **downconverter**. The receiver is also known as the 'front end'.
[See also **receive chain**.]

RECONNAISSANCE SATELLITE

A military satellite, usually in a **low Earth orbit**, concerned primarily with high **resolution** imaging of 'enemy territory'. Colloquially termed a 'spy satellite'.
[See also **remote sensing**.]

RECOVERY

The retrieval of a space **capsule**, recoverable **solid rocket booster** or any other item, typically from a body of water, following a **splashdown**.

RECYCLE COUNTDOWN

The procedure whereby a countdown clock is re-set to an earlier point to continue the **countdown**, usually following the solution of a problem which led to a **hold** in that countdown. The procedure is only followed when the hold was unforeseen; in the case of a **built-in hold** the countdown continues from the point at which it was halted.

REDSTONE

An American single-**stage** liquid **propellant** medium range ballistic missile (MRBM) developed by the US Army in the 1950s (**oxidiser**: **liquid oxygen**; **fuel**: 75% ethyl alcohol, 25% water). Used as a **launch vehicle** for the 15-minute **sub-orbital** flights of the **Mercury** programme (the **Atlas** booster was used for the orbital flights). No

longer **operational**.
[See also **Jupiter**, **Thor**.]

REDUNDANCY

An attribute of a **system** in which a 'spare' or **back-up** device is
available in case of failure of the **prime system**. If a system is to be 'fully
redundant', all parts of the prime system must be duplicated in the
back-up system: this is known as 'two-for-one redundancy'. It is
unusual to duplicate all the devices in, for example, a spacecraft
communications payload due to the restriction on mass, but
redundancy is typically provided for critical components (thus two
amplifiers connected in parallel provide 'two-for-one redundancy'; a
'redundancy scheme' where one spare is provided for each pair of
amplifiers is 'three-for-two redundancy', and so on).
[See also **single-point failure**.]

REDUNDANT

Spare: a redundant component, **system**, etc, is one designed to replace
the **prime system** in case of failure; otherwise termed a **back-up**. See
redundancy.

RE-ENTRY

The return of a **spacecraft** or other man-made body into the Earth's
atmosphere. For example, the American **Space Shuttle** re-entry is
considered to occur at an altitude at 122 km (400 000 feet).
[See also **entry interface**, **atmospheric drag**, **plasma sheath**, **thermal
protection system**, **ablation**.]

RE-ENTRY CORRIDOR

A volume of space through the Earth's atmosphere within which a
returning **spacecraft** must pass to ensure a safe **re-entry** [figure R2]. If
the angle of the re-entry **trajectory** relative to the Earth's surface is too
shallow, the spacecraft may 'skip off' the atmosphere (like a stone
skipping off a water surface); if the angle is too steep, it will burn up due
to excessive frictional heating (e.g. the nominal angle for the **Apollo**
command module was 6° $\pm$ 1°, and the corridor was 42 km wide).

REFLECTOR ANTENNA

See **antenna**.

REFRACTORY METAL

A metal able to withstand high temperatures without fusion or

decomposition, typically with a melting point above 2200 °C (e.g. columbium, molybdenum, tantalum and tungsten).
[See also **ceramics**.]

REGENERATIVE COOLING

The practice of pumping **liquid propellant** through the walls of exhaust **nozzle**s, and sometimes **combustion chamber**s, before it enters the chamber itself [see frontispiece]. Cooling the nozzle means that it can withstand higher exhaust-gas temperatures (which can lead to an increase in **specific impulse**) without recourse to increasingly exotic heat-resistant materials. **Cryogenic propellant**s are particularly useful in this regard; regenerative cooling is clearly impossible with **solid propellant** motors.
[See also **rocket engine**.]

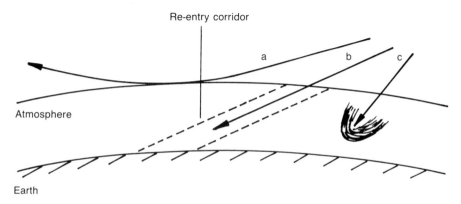

Figure R2 A spacecraft **re-entry corridor**. a, angle too shallow—vehicle skips off atmosphere; b, correct angle; c, angle too steep—vehicle burns up.

REGULATED POWER SUPPLY

A spacecraft **power supply** which provides a constant voltage, typically to within plus or minus 1% of the nominal value, whether the spacecraft is in sunlight or **eclipse**, whether relying on its primary or **secondary power supply**.

Satellite power subsystems, for instance, can be fully regulated; totally unregulated; or regulated in sunlight, when power can be derived from the **solar array**, and unregulated in eclipse, when the output voltage falls to that provided by the **battery**.
[See also **unregulated power supply**.]

RELIABILITY

A branch of spacecraft engineering which quantifies the likelihood of a failure at component, device, **subsystem**, **system** or **spacecraft** level and provides feedback to the design teams to reduce the probability of failure. Apart from improved standards of quality control, material and component selection, handling and testing, reliability is improved by incorporating the concept of **redundancy**, thereby minimising the possibility of **single-point failure (SPF)**.

RELIEF VALVE

See **vent and relief valve**.

RE-LIGHT

See **engine re-start (capability)**.

REMOTE MANIPULATOR SYSTEM (RMS)

In general, a jointed, mechanical arm (and its associated control systems) attached to a spacecraft and used, by remote control, to handle equipment, deploy and retrieve **payload**s, etc. In particular, the mechanical arm mounted on the side of the **payload bay** of the American **Space Shuttle** [see figure R3]. The Shuttle manipulator arm is nearly 16 m in length and 0.38 m in diameter, and is capable of handling payloads of nearly 30 tonnes. It comprises two **boom**s and an **end effector** (hand) with shoulder, elbow and wrist joints, which give it six **degree**s **of freedom**. It is operated from the aft **flight deck** of the Shuttle, which overlooks the payload bay.
[See also **telepresence**.]

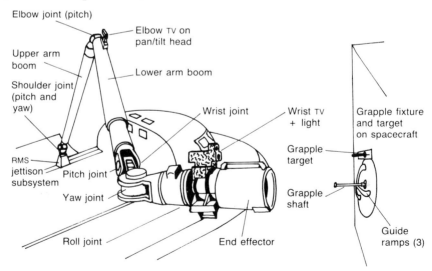

Figure R3 The **remote manipulator system** (RMS) used on the American **Space Shuttle**.

REMOTE SENSING

Detection and **data** collection from a distance.

Remote sensing satellites are usually placed in high-inclination, near-**polar orbit**s, which give full Earth coverage over a period of days with a regular 're-visit capability' to enable data to be updated. Typical **payload**s allow observations in a variety of wavebands from the ultra-violet to the infrared, with a multitude of applications, including the determination of crop health, pollution detection, map-making and 'intelligence gathering'. In fact, the only practical difference between the remote sensing satellite and the military 'spy satellite' is the **resolution** of the **sensor**s, which is generally greater in the latter case. [See also **meteorological satellite, radar.**]

RENDEZVOUS

A meeting between two or more **spacecraft** engineered by a matching of **orbital parameter**s; usually a preamble to **docking**. The process is not as straightforward as it may at first appear. The radius of a spacecraft's **orbit** is dependent upon its orbital velocity, such that a spacecraft in a higher orbit has a longer **orbital period**. This means that, for two spacecraft travelling one behind the other in the same orbit, the one behind must decrease its orbital height to reduce its period and 'overtake' the other. Simply increasing its orbital velocity would add energy to the orbit and increase its height; its period would increase and it would lose ground on the first spacecraft. Therefore its orbital velocity/energy must be decreased: height decreases, period decreases and it overtakes. To complete the rendezvous, it must regain its orbital height from a position in front of and below the first spacecraft, which it does by increasing its orbital velocity/energy in a carefully controlled way.
[See also **orbital control, equilibrium point.**]

REPEATER

A device which receives, amplifies and retransmits electrical signals (e.g. satellite repeater, communications repeater). See **transponder**.

RESIDUAL OXYGEN

Oxygen atoms in **low Earth orbit** which can degrade certain spacecraft **surface coatings** by impact [see **degradation**].

RESIDUALS

Jargon: the **propellant**s or **pressurant**s left in a tank at the end of a **mission**. It is usually not possible to dispense the entire contents of a propellant tank and a certain volume of propellant will remain. Since the **lifetime** of many **unmanned spacecraft** is governed by the amount of **station-keeping** propellant carried, it is important to make allowances for 'residuals': an entry for such should therefore appear in a spacecraft's **propellant budget**.

RESOLUTION

A measurement of the ability of an instrument, a recording medium or an observer to 'resolve' or distinguish between separate parts of an **image**; also called 'resolving power'.
[See also **pixel**.]

RESTRAINT

Any device on or in a **spacecraft** which restricts an **astronaut**'s movement, especially important in the **microgravity** environment. For example 'foot restraints' (used by astronauts on **extra-vehicular activity** (EVA)); and 'sleep restraints' (sleeping bags used on the American **Space Shuttle**).

RETROFIT

To add components or **systems** after manufacture, often necessary when updating equipment or adding safety systems, etc.

RETROGRADE ORBIT

An **orbit** in which the orbiting body has the opposite sense of rotation as the orbited body; the opposite of **prograde orbit**.

RETRO-PACK

In general, a 'package' of **retro-rockets** used to decelerate a spacecraft prior to **re-entry**; in particular, the package of **solid propellant** rocket motors mounted over the **heat shield** of the manned **Mercury** capsule, jettisoned after use.

RETRO-REFLECTOR

See **laser ranging retro-reflector**.

RETRO-ROCKET

A small, auxiliary rocket engine on a **launch vehicle** or **Tpacecraft** that produces **thrust** in the opposite direction to the direction of **flight** in order to decelerate the vehicle. Often abbreviated to 'retro'.

As examples, most launch vehicle **stages** have retro-rockets which fire

after **stage separation** to prevent the jettisoned stage colliding with the upper stage before its engines are ignited; spacecraft returning to Earth or landing on another **planetary body** sometimes use retro-rockets to make a 'soft landing'. Most retro-rockets use **liquid propellant**s, but **solid propellant**s have also been used [see **retro-pack**].
[See also **rocket engine, rocket motor**.]

REUSABLE LAUNCH VEHICLE

A **launch vehicle** which can be used more than once; the alternative to an **expendable launch vehicle**.

In most cases only part of the vehicle is truly reusable: for instance, the American **Space Shuttle** has a reusable **orbiter** and an expendable **external tank** (ET). Its **solid rocket booster**s (SRBs) can be refurbished for later use.

RE-VISIT CAPABILITY

See **remote sensing**.

RF

See **radio frequency**.

RF CHAMBER

See **anechoic chamber**.

RF POWER

Power supplied by devices operating at **radio frequency** (RF). See **power**.
[See also **radiated power, power flux density, saturated output power, power supply, power bus, power budget**.]

RF SENSOR

Radio-frequency sensor: a spacecraft **sensor** used to maintain accurate **pointing** of a spacecraft's **antenna**s. The sensor detects an RF signal transmitted from a known point in the **service area** and an **antenna pointing mechanism** drives the antenna to maintain the maximum signal level. The most common type is the monopulse RF sensor.
[See also **earth sensor**.]

RF WAVE

An abbreviation for **radio-frequency** wave: any waveform in the **electromagnetic spectrum** which can be classed as 'RF', i.e. between about 10 kHz and 300 GHz. See **radio frequency (RF)**.

RFNA

An acronym for red fuming nitric acid, a substance commonly used in the past as a rocket **propellant**.
[See also **liquid propellant**.]

RHCP (RIGHT-HAND CIRCULAR POLARISATION)

See **polarisation**.

RMS

An acronym for **remote manipulator system**.

ROBOTICS

A branch of engineering concerned with automated machines designed and programmed to perform specific mechanical functions with or without human intervention (e.g. a planetary **probe** like the **Viking** lander designed to sample the surface of Mars). When a mechanical manipulator is controlled entirely by human intervention from a distance, the technique is known as **telepresence**.

ROCKET

(i) A propulsive device using a mixture of **propellant**s (**fuel** and **oxidiser**) which, upon **ignition**, provide a reaction force to propel a space vehicle, missile, rocket-assisted aircraft, etc.

(ii) A colloquial term for **launch vehicle**.

[See also **rocket engine, rocket motor, retro-rocket, booster, reaction control thruster**.]

ROCKET BOOSTER

See **booster**.

ROCKET ENGINE

A propulsive device in which **liquid propellant**s are burnt in a combustion chamber to provide a reaction force to propel a vehicle; the part of a chemical **propulsion system** in which propellants are burned and the combustion products are used to produce **thrust** [see figure R4 and frontispiece]. Although the terms engine and motor are generally interchangeable, it is customary for propulsion devices using liquid propellant to be called 'engines' and those using **solid propellant** to be called 'motors'.

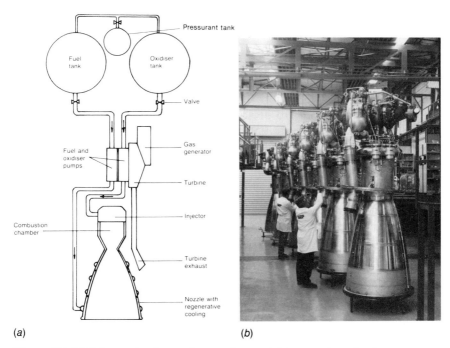

(a)   (b)

Figure R4 (a) Schematic of a **rocket engine** and (b) a number of Viking engines for the **Ariane** launch vehicle's first stage (from top to bottom: **turbopump** assembly with **exhaust** on left, **combustion chamber**, **nozzle**). [SEP]

A typical liquid rocket engine comprises a propellant delivery and **injection** system, an **ignition system**, a **combustion chamber** and an exhaust **nozzle**. Liquid propellant engines are either pump-fed or pressure-fed: although the former are more complex, the latter require thicker-walled propellant tanks to withstand the pressure. In general, the liquid engine is more complex than the solid motor: it requires a network of pipes and pumps (and/or **pressurant** tanks) to deliver the propellant to the combustion chamber, whereas a 'charge' of solid propellant makes its own 'chamber'. In operation, however, the liquid engine offers a far greater degree of control: it can be stopped in-flight and re-started, the flow of propellant can be varied, and the combustion chamber/nozzle assembly can be gimballed to vary the direction of the **thrust vector**, thereby steering the vehicle.

In addition, the nozzle and combustion chamber walls can be cooled by piping unburnt propellant through them (**regenerative cooling**). Since solids do not have this advantage, their **burn** times are limited by the thermal properties of the **materials** used in the nozzle.

[See also **rocket motor, turbopump, gimbal, liquid apogee engine (LAE), ascent engine, descent engine, combined propulsion system, propellant.**]

ROCKET MOTOR

A propulsive device in which **solid propellant**s are burnt to provide a reaction force, or **thrust**, to propel a vehicle. Also called a 'solid rocket motor' (SRM). Although the terms motor and engine are generally interchangeable, it is customary for propulsion devices using solid propellant to be called 'motors' and those using **liquid propellant** to be called 'engines'.

A typical solid rocket motor comprises only a few major components: a **motor case** which contains the **propellant grain**, a surrounding insulating blanket or **propellant liner**, an exhaust **nozzle** and an **ignition system** [see figures R5 and A5]. The motor case is typically made from a carbon or **Kevlar composite** by a process called **filament winding**, which produces a strong yet light-weight structure. The propellant liner acts as both a **thermal blanket** and a **flame inhibitor**, and also supports the propellant grain during manufacture, as it is poured into the open-ended motor case. In the terminology of liquid propulsion systems, the motor case is both the **propellant tank** and the **combustion chamber**.

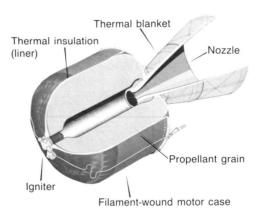

Thermal blanket

Thermal insulation (liner)

Nozzle

Propellant grain

Igniter

Filament-wound motor case

Figure R5 Cutaway of a MAGE 2 **rocket motor** used as a satellite **apogee kick motor** (AKM). [SEP]

In operation, the solid propellant motor has no need of the complex piping, pumps and **pressurisation** systems of the liquid **rocket engine**: it is like comparing a safety match to a gas-lighter. It does, however, have to be designed to give precisely the required amount of thrust and total **impulse**, since it cannot be 'throttled', easily stopped or re-started.

The only way to control the magnitude of the thrust is to design the propellant charge to offer varying surface areas for combustion throughout the 'burn' [see **propellant grain**]. Thrust vectoring is more difficult with a solid motor than with a liquid engine since only the nozzle can be gimballed, but systems using vanes or injecting inert gas into the nozzle to deflect the **plume** have been used.
[See also **solid rocket booster, reaction control thruster, thrust vector, gimbal, throttling**.]

ROCKET NOZZLE

See **nozzle**.

ROCKET STAGE

See **stage**.

ROLL

A rotation about the roll axis—see **spacecraft axes**.
[See also **roll programme, barbecue roll, dutch roll**.]

ROLL AXIS

See **spacecraft axes**.

ROLL PROGRAMME

The sequence of commands to a **launch vehicle** which cause it to rotate about its vertical axis (**roll**) to provide gyroscopic stability. The roll programme typically begins a short time after **lift-off** and continues until **payload** separation, although the rate of roll may be reduced prior to separation. The roll programme of the American **Space Shuttle** differs in that it is a single roll **manoeuvre** (clockwise rotation about the vertical axis) of a number of degrees (e.g. 120,140) dependent upon the required **orbital inclination**.
[See also **spin-stabilised**.]

ROM

An acronym for read only memory, a computer memory device to which the user cannot add information (as opposed to a **RAM**). Variants are PROM (programmable read only memory) and EPROM (erasable programmable read only memory).

ROTATING SERVICE STRUCTURE (RSS)

A **service structure**, attached to the 'fixed service structure' at the **launch pad**s used by the American **Space Shuttle**, which can be rotated to provide access to the Shuttle's **payload bay**, particularly for loading [see figure S3]. In a typical loading operation, a **payload canister** is delivered to the pad, and lifted into the 'payload changeout room' on the RSS. The RSS is then rotated towards the **orbiter** until the changeout room doors form a seal with the open payload bay, and the payloads are transferred to the orbiter.

ROTATIONAL VELOCITY (of EARTH)

See **plane change**.

RP-1

See **kerosene**.

RTG

See **radioisotope thermoelectric generator**.

RUNWAY (for SPACE SHUTTLE)

See **shuttle landing facility (SLF)**.

S

S-BAND

2–4 GHz—see **frequency bands**.

S-BAND BLACKOUT

A period during the **re-entry** of a **spacecraft** during which radio communications are impossible due to the **ionisation** of the atmosphere surrounding the vehicle (e.g. its duration for the American **Space Shuttle** is about 15 minutes). 'S-band' refers to the **frequency** in common use for radio communication between the ground and a **manned spacecraft** [see **frequency bands**].

SADA

An acronym for solar array drive assembly. See **solar array**.

SADE

An acronym for solar array drive electronics. See **solar array**.

SADM (SOLAR ARRAY DRIVE MECHANISM)

See **BAPTA**.

SAFING

The process whereby a space vehicle is 'made safe' following a return to Earth. It includes the removal of hazardous residual **propellant**s, the draining and purging of propellant feed-lines, and the removal of explosive devices such as pyrotechnic **actuator**s.
[See also **purge**.]

SALYUT

A series of Soviet **space station**s launched and operated throughout the 1970s and 1980s. Salyut 1, which was launched in April 1971, was followed by two failures (Salyut 2 and a Salyut-type vehicle Cosmos 557) and the successful flights of Salyuts 3–7: 3 and 5 for military applications; 4 and 6 mainly civil. Salyut 7 combined both applications and was significantly redesigned. It had a total habitable volume of about 100 m³ and comprised three **modules**: a forward transfer compartment (basically an access tunnel used for **EVA**s); an operations compartment used as the main living and working area; and a rear transfer compartment to allow connection with visiting **Soyuz** 'ferries' and **Progress** re-supply 'tankers'. After three **crew**s had occupied Salyut 7 for periods of 211, 150 and 237 days, with short stays by visiting crews, the station was boosted into a higher **orbit** in 1986 and left for possible future experimental use.
[See also **Mir**, **Skylab**.]

SAN MARCO PLATFORM

A **launch platform** in the Indian Ocean 5 km from the Kenyan coast (at approximately 3°S, 40°E) operated by the Aerospace Research Centre of the University of Rome. Launches are controlled from the adjacent Santa Rita control platform. The facility was developed in the mid 1960s to launch the Italian San Marco satellite using NASA's **Scout** launch vehicle.

SANGER

A two stage **aerospace vehicle** proposed by West Germany as a next-generation satellite launcher. Named after Eugen Sanger (1905–1964) who developed the original concept in 1942. The proposed second stage can be either a manned **space shuttle** (called Horus), or an unmanned cargo transporter (Cargus).
[See also **HOTOL**.]

SATELLITE

(i) A **celestial body** in orbit around another celestial body.

(ii) A **spacecraft** in orbit around a celestial body, such as the Earth, the Sun or any other of its planets and moons. Although the terms 'spacecraft' and 'satellite' are often used synonomously, a spacecraft ceases to be a satellite when it leaves **orbit**.

SATELLITE COMMUNICATIONS

The collective term for information transmitted via a **communications**

satellite. Also called satellite telecommunications.
[See also **communications, telecommunications**.]

SATELLITE SWITCHING

The routing or distribution of messages, **data**, etc, performed on board a **satellite** rather than on the ground. This is a relatively advanced concept in which the satellite is described lyrically as 'the switchboard in the sky'. Most satellite **transponders** operate simply in that they receive an **uplink** signal, amplify and filter it, change its frequency and then transmit it into a single **downlink** beam. For major trunk telecommunications links, the downlinked signal is received by a single **gateway station** and routed from there into the terrestrial network. With satellite switching, the uplinked signal is labelled with its destination and the satellite routes it into one of several **beams** aimed at several earth stations within the **coverage area**.
[See also **multiple beam, satellite-switched TDMA**.]

SATELLITE-SWITCHED TIME DIVISION MULTIPLE ACCESS (SSTDMA)

A version of TDMA where the switching is done on board the **satellite** rather than at the **earth station**. See **satellite switching** and **time division multiple access (TDMA)**.

SATURATED OUTPUT POWER

The output power of an **amplifier** driven at **saturation**; the maximum output power available from a **high-power amplifier**. Beyond saturation, any increase in the input power results in a decrease in output power.
[See also **back-off (from saturation)**.]

SATURATION

The condition of a **high-power amplifier** when driven to its maximum output: it is said to be 'at saturation'.
[See also **saturated output power, back-off**.]

SATURN 1B

An American **launch vehicle** developed in the early 1960s for the Earth orbit tests in the **Apollo** lunar programme; no longer **operational** [see figure S1]. It comprised two **stages**: the first stage, called the S-1B, used the propellants **liquid oxygen** and **kerosene**; the second stage (S-IVB) used liquid oxygen/**liquid hydrogen**.

The Saturn 1 booster, formerly known as the Saturn C-1, had a first stage developed from the early American single-stage rockets: it was effectively eight **propellant tank**s from the **Redstone** clustered round a central tank based on the **Jupiter**. The Saturn 1B was an **uprate**d Saturn 1, with the third stage developed for the **Saturn V** used as a second stage to allow it to orbit the Apollo spacecraft. The Saturn 1B was first launched in February 1966 with an unmanned Apollo. It was also used in October 1968 to launch the first manned Apollo (Apollo 7), in 1973 to launch three crews to **Skylab**, and in 1975 to launch the crew for the **Apollo–Soyuz test project (ASTP)**.

Figure S1 Launch of a **Saturn 1B** rocket carrying an **Apollo** command and service module on the **Apollo–Soyuz test project** mission in July 1975. Since the launch tower was designed for the larger **Saturn V**, the Saturn 1B had to be launched from an elevated platform (dubbed the 'milkstool'). [NASA]

SATURN V.

An American **launch vehicle** developed in the 1960s for the **Apollo** lunar programme; no longer **operational**. Formerly known as the Saturn C-5, it comprised three **stage**s: the first stage, called the S-1C, used the propellants **liquid oxygen** and **kerosene**; the second and third stages (S-II and S-IVB) used liquid oxygen/**liquid hydrogen**. The Saturn V was first launched 9 November 1967 for the unmanned Apollo 4 test

launch, and last used 14 May 1973 to launch the American **Skylab** space station.
[See also **Saturn 1B, specific impulse.**]

SCHMIDT CAMERA

See **Schmidt telescope**.

SCHMIDT TELESCOPE

A specialised telescope for astronomical photography, devised by the Estonian optician Bernhard Voldemar Schmidt (1879–1955). Also called a Schmidt camera. Most telescope optics are unsuitable for photography because of the distortions they cause to objects far from the centre of the field of view. This means that high magnification and a large field of view are contradictory requirements. To enable undistorted photography of large star-fields, Schmidt devised, in 1930, an optical 'corrector plate' for use in a telescope with a spherical primary mirror. The light passes through the plate, which corrects for the spherical aberration of the primary, and is reflected from the primary onto a photographic plate mounted at its focus [see figure N1]. [See also **space telescope, Newtonian telescope, Cassegrain reflector, Gregorian reflector, Coudé focus.**]

SCOUT

An American four-stage **solid propellant** launch vehicle developed in the late 1950s to launch relatively small, typically scientific, **payload**s into **low Earth orbit** (LEO). It became **operational** in 1960 and its derivatives are still in use. It can be launched from the Western Test Range at **Vandenberg Air Force Base** (VAFB), Wallops Island, Virginia or the **San Marco platform**. An uprated third **stage** provides a payload capability of about 220 kg to LEO (550 km).
[See also **Atlas, Delta, Titan.**]

SCPC

See **single channel per carrier**.

SCRAMBLER

An electronic device used to make radio transmissions unintelligible to unauthorised receivers. See **scrambling**.

SCRAMBLING

A process by which radio transmissions are 'mixed up' to make them unintelligible to unauthorised receivers, used particularly for **cable TV** and TV broadcast direct from **satellite**s (DBS). Also used in telephone systems, etc. Typically, **baseband** signals are scrambled by dividing them into a number of frequency sub-bands and transposing them in a predetermined manner. The authorised receiver has the 'key' to the transposition and can unscramble the signal.

SCRUB

American slang for postponement of a **launch** (e.g. 'the launch has been scrubbed').

SDI

See **strategic defence initiative**.

SECOND SPACE VELOCITY

See **escape velocity**.

SECOND-SURFACE MIRROR (SSM)

A thin sheet of glass or quartz, silvered or aluminised on one side, bonded to the exterior surface of a **spacecraft** as part of the **thermal control subsystem** [figure S2].

Figure S2 The ECS-1 communications satellite during **solar array** deployment tests. Note the highly reflective **second-surface mirror**s (SSMs). [ESA].

The term 'second surface' is derived from its dual function as a thermal emitter and **solar reflector**: glass is an excellent emitter over the infrared spectrum, so thermal energy from the spacecraft can be conducted to the SSMs and radiated into space (from the outer or 'first surface' of the mirror); moreover, since glass is transparent over most of the **solar spectrum**, the majority of the incoming solar radiation reaches, and is reflected from, the coated rear surface (the 'second surface').

A typical SSM has a substrate based on a borosilicate glass doped with cerium dioxide, similar to solar cell **coverslip**s. SSMs are not easy to fit to anything but flat surfaces since they are essentially rigid, but their high resistance to electron and UV irradiation has led to their widespread use.
[See also **optical solar reflector**.]

SECONDARY POWER SUPPLY

The secondary source of **DC power** available aboard a spacecraft. For most spacecraft it is the **battery**, but some manned space vehicles, which use the **fuel cell** as the primary source, may use the **solar array** as their secondary source. There are no rules which place a power source in one category or the other.
[See also **primary power supply**.]

SEE

See **single-event effect**.

SEISMOMETER

A device for measuring the strength of an earthquake, moonquake or similar ground tremors on any other **planetary body**; also called a 'seismograph'.

SEMI-MAJOR AXIS

In astronomy, one of the halves of the **major axis** of an **orbit**.

SENSOR

Any device that receives a signal or stimulus and responds to it with a proportional signal which can be used in a **spacecraft** subsystem. It may be part of a **subsystem** which controls **pointing** [see **sun sensor**, **earth sensor**, **star sensor**, **RF sensor**], or a **detector** in the **payload** of a **remote sensing** or **astronomical satellite**.
[See also **attitude control**, **space astronomy**.]

SEQUENCER

See **countdown sequencer**.

SERVICE AREA

An area on the surface of the Earth, defined for the purposes of **satellite communications**, within which the administration originating a service can demand protection against **interference** from other transmissions [see figure C4].

A **receiver** in the service area should therefore receive an interference-free signal from that country's satellite. This is one of the rulings of the World Administrative Radio Conference (**WARC**) designed to create a regime where, ideally, no telecommunications service interferes with any other. It is assumed that the service area lies inside the political boundary of the originating country, but WARC is not specific on that point, recognising that some political boundaries are subject to dispute.

[See also **coverage area, beam area, protection ratio**.]

SERVICE MODULE

A self-contained section of a modular spacecraft containing the subsystems which support the **payload**, whether 'man or machine' [see figures M2 and M3].

The service module of a **communications satellite**, for example, contains the subsystems that support the **communications payload** (e.g. **power, thermal** and **attitude and orbital control system**s). The term may also include the spacecraft's propulsion system or there may be a separate **propulsion module**, depending on the design. The other main modules are the **payload module** and **antenna module**. The chief advantage of the modular design concept is that the modules can be assembled and tested separately (by different contractors in different countries if required) without heed to the constraints of the other modules. Final **integration** and testing follows at a later stage.

[See also **Apollo**.]

SERVICE STRUCTURE

A tower framework which allows access to a **launch vehicle** on its launch pad for maintenance, fuelling, **payload** loading, etc [see figure S3]. Also called an access tower, service tower, umbilical tower or gantry.

A service structure can be fixed to a **launch pad** or **launch platform** (a 'fixed service structure'), or mounted on rails to enable it to be moved away from the vehicle for **launch** (a 'mobile service structure').

[See also **rotating service structure, swing-arm, access arm**.]

SERVICE TOWER

See **service structure**.

Figure S3 Space Shuttle Atlantis at **launch pad** 39B, **Kennedy Space Center**. This photograph shows the fixed **service structure** with **swing-arm**s carrying the **beanie cap** (near top) and **white room** (centre); the **rotating service structure** (at left) with its 'payload changeout room' which fits over the open **payload bay**; and the **launch platform** with its **tail service mast**s (below the **orbiter**'s wings). [NASA]

SEU

An acronym for single-event upset. See **single-event effect**.

SHAKE TABLE

See **vibration table**.

SHEAR

A form of deformation or fracture in which parallel planes of a body slide over one another; in physics, the deformation of a body expressed as the lateral displacement between two points in parallel planes divided by the distance between the planes.
[See also **shear web**, **wind shear**.]

SHEAR WEB

A structural element which transfers or shares structural (**shear**) loads

between panels or other components.
[See also **load path, doubler**.]

SHF (SUPER HIGH FREQUENCY)

See **frequency bands**.

SHIP EARTH STATION

An **earth station** in the maritime **mobile-satellite service (MSS)** located onboard a vessel (approval for land-based use is also attainable).
[See also **coast earth station**.]

SHIRT-SLEEVE ENVIRONMENT

An environment which is comfortable for someone wearing typical indoor clothing (literally, 'in their shirt-sleeves'); the environment intended for the **cabin** of a **manned spacecraft**, within which the wearing of any type of **spacesuit** is unnecessary.
[See also **environmental control and life support system (ECLSS)**.]

SHOCK DIAMONDS

A pattern of shock waves (a number of linked diamonds) sometimes visible in a rocket exhaust.

SHOCK SUPPRESSION SYSTEM

See **sound suppression system**.

SHOT NOISE

A source of **noise** in a communications system which is produced by discrete charges in an electric current. It is therefore impulsive and random, as opposed to the continuum of noise as represented by **thermal noise**. For example, shot noise may be produced by the electrons in a thermionic diode, a **klystron** or a **travelling wave tube** (TWT).

SHROUD

A protective **fairing** which houses the **payload** of a **launch vehicle**. More fully known as a 'payload shroud' or 'launch shroud' [see figure S4]. The shroud protects the payload from the environmental characteristics of the atmosphere prior to **launch**, and frictional heating and other dynamic effects during its passage through the atmosphere. Most shrouds are divided pyrotechnically into two equal halves and jettisoned once the vehicle is above the denser part of the atmosphere.
[See also **SYLDA, SPELDA, SPELTRA, pyrotechnics, payload envelope**.]

SHUNT DUMP REGULATOR

A device which takes excess spacecraft **power** from the **solar arrays**, converts it to heat by dissipation in a bank of resistors and then 'dumps' or radiates it into space. Alternative names for units with the same or similar capabilities are 'shunt voltage limiter' and 'shunt voltage regulator assembly'.

An alternative to the shunt dump regulator is the array shunt regulator, which connects or disconnects sections of the array in accordance with demand. Since power is only drawn into the spacecraft when it is needed, thermal dissipation is kept to a minimum. [See also **heat rejection**.]

Figure S4 The OTS-2 communications satellite in the payload **shroud** of its **Delta** launch vehicle. [ESA]

SHUTTLE

See **Space Shuttle**.

SHUTTLE CARRIER AIRCRAFT (SCA)

The specially adapted Boeing 747 used to ferry American Space Shuttle **orbiter**s from place to place, typically from the landing site at **Edwards Air Force Base** to the **launch site** at **Kennedy Space Center**. It would

also be used in the eventuality of a Shuttle landing at one of the 'emergency runways' in other parts of the world. The SCA was formerly used to carry the orbiter **Enterprise** in the approach and landing test (ALT) programme.
[See also **mate/de-mate device**.]

SHUTTLE LANDING FACILITY (SLF)

The formal name for the runway at **Kennedy Space Center** on which American Space Shuttle **orbiters** land. It is 4575 m (15 000 ft) in length with a 305 m (1000 ft) over-run at each end, 91 m (300 ft) wide and oriented northwest–southeast. A similar facility exists at **Edwards Air Force Base** in California, except that it has an 8 km over-run which extends into a dry lake-bed. Orbiters are guided automatically to a landing by a microwave landing system (MLS) which measures the vehicle's position relative to the runway centre-line and its elevation angle, and instructs it to make alterations to its **glide path** as necessary.

SIDELOBE

See **antenna radiation pattern**.

SIDEREAL PERIOD

An orbital period defined in the **frame of reference** of the 'fixed stars'. See **orbit**.

SIGNAL

A variable parameter, such as a current or electromagnetic wave, used to convey information through an electronic or microwave circuit, or through **free space**. When the signal is the **output** of an item of equipment such as a telephone, telex, TV camera, computer, etc, it may also be referred to as a **baseband** signal. For transmission via a communications system, the signal is modulated (or superimposed) onto a **radio-frequency** carrier wave.
[See also **signal-to-noise ratio**, **carrier**, **modulation**.]

SIGNAL-TO-NOISE RATIO

The ratio of the power level of a **signal** to the power level of the **noise** at the same time and place in a circuit or any signal-carrying medium. Abbreviated to S/N or, less frequently, SNR, this quantity refers to a signal at **baseband** as opposed to RF or IF. S/N may also be measured using other parameters, such as the amplitude of a wave displayed on an oscilloscope tube, but these other parameters are simply a convenient illustration of signal and noise power. In practical terms, the higher the S/N, the less the **interference** with reception.

[See also **noise power**, **carrier-to-noise ratio**, **radio frequency**, **intermediate frequency**.]

SIMULATION

A reproduction of specific conditions for the purposes of testing or training. See **simulator**.

SIMULATION HEATER

A **heater** designed to simulate a deactivated device which produces heat when it is operating; also called a substitution heater.

For example, if a **TWTA** is switched off, perhaps during an **eclipse**, its temperature drops and the **radiator** it is attached to cools. This creates an undesirable temperature gradient which upsets the **thermal balance** of the spacecraft. To stop this happening, TWTA simulation heaters, which replace at least some of the heat produced by the amplifier, are bonded to the radiator or adjacent structure.
[See also **line heater, thermal control subsystem**.]

SIMULATOR

Any device or facility which reproduces specific conditions for the purposes of testing or training. Usually applied to devices which simulate conditions aboard a **manned spacecraft**, especially with regard to its control in flight (i.e. a spacecraft simulator), but extended to ground-based control and operations facilities.
[See also **ground support equipment, centrifuge, thermal-vacuum chamber, acoustic test chamber, vibration facility, anechoic chamber**.]

SINGLE CHANNEL PER CARRIER (SCPC)

A derivative of the FDMA multiplexing systems in which the transponder **bandwidth** is subdivided so that each **baseband** channel is allocated a separate **transponder** subdivision and an individual **carrier**.
[See also **frequency division multiple access (FDMA)**.]

SINGLE CONVERSION TRANSPONDER

See **upconverter**.

SINGLE-ENTRY INTERFERENCE

A term used in connection with **communications systems** to indicate the presence of only one interfering signal. See **protection ratio**.

SINGLE-EVENT EFFECT (SEE)

A term applied to damage caused by the random effects of natural

radiation sources on **spacecraft** electronic components, particularly semiconductor storage devices ('memory chips'); also called 'single-event upset' (SEU). The effects are primarily due to particles generated in **solar flare**s and **cosmic radiation** (e.g. protons, neutrons, pions, muons and heavy ions) which cause a 'change of state' or 'polarity reversal' in the device. The particle deposits a charge and causes a transient pulse which can be decoded as an instruction (to turn it on or off, etc). The increasing importance of SEE is partly due to the ever-decreasing size of chip features (the critical feature size is about 3 microns), but also due to the decrease in the electrical charge required to activate the device, which is becoming closer to that deposited by the particles themselves. SEE is more pronounced for satellites in **polar orbit**s, since they lose the protection of the Earth's magnetic field over the polar regions. Although **data** are lost while the memory is reloaded, SEE is not too much of a problem in Earth orbit, but it could prove more serious for the more autonomous interplanetary spacecraft.
[See also **hardening**.]

SINGLE-EVENT UPSET (SEU)

See **single-event effect (SEE)**.

SINGLE HOP

A signal route in a **communications satellite** system which includes a single pass through a **satellite**: i.e. there is one **uplink** path and one **downlink**, the remainder of the route being terrestrial. A route which includes two satellites is known as a **double hop**.
[See also **terrestrial tail**.]

SINGLE-POINT FAILURE (SPF)

A concept in **reliability** engineering whereby a single component can cause the failure of a complete **system** because there is no **redundancy** provision (i.e. no spare component or '**back-up**'). For example, although a satellite **communications payload** typically includes **redundant** amplifiers, it outputs to a single **antenna**. If any part of the antenna fails or the reflector itself fails to **deploy**, the section of the **payload** to which it is attached will be unable to function. The antenna, or the antenna deployment mechanism where it exists, represents a potential 'single-point failure'. Space-based systems are designed to keep SPFs to an absolute minimum. Other SPF examples include a satellite **apogee kick motor** (AKM), the main mirror of the Hubble **space telescope** and the nose-wheel of an American Space Shuttle **orbiter**.
[See also **deployable antenna**.]

SINGLE STAGE TO ORBIT (SSTO)

The concept of a **launch vehicle** capable of delivering a **payload** to **orbit** (usually **low Earth orbit**) using only one rocket **stage**. Although several designs have been postulated, there is unlikely to be an **operational** SSTO before the late 1990s. The favoured designs fit into the category of the **aerospace vehicle**, with both **rocket engine**s and **lifting surface**s (e.g. HOTOL), as opposed to the conventional vertical take-off launch vehicle.

SKIN

The outermost layer of a **spacecraft** or **launch vehicle** structure. Typically a thin metal covering designed to help protect a spacecraft from the high temperatures of **re-entry** (high 'skin temperature'); or a lightweight covering on a rocket **stage** supported by **longeron**s and **stringer**s.
[See also **face-skin**, **airframe**, **fairing**, **thermal protection system**, **heat shield**, **aerodynamic heating**.]

SKIRT

(i) A structural extension joining two sections of a **spacecraft** or **launch vehicle**, usually flared to join two sections of different diameter (e.g. adapter skirt, skirt assembly, etc).

(ii) A **fairing** surrounding the **engine bay** at the base of a launch vehicle's first **stage**. Also called a 'cowling', particularly when it is a fairing round an individual engine.

[See also **interstage**.]

SKYLAB

An American **space station** designed for terrestrial and astronomical observation, and research into human capabilities in and reaction to the space environment [see figure S5]. Skylab was launched on 14 May 1973 by a **Saturn V** into a 440 × 427 km, 50° inclination **orbit** and visited by three **crew**s for period of 28, 59 and 84 days, until it was vacated in February 1974. The crews were transported by **Saturn 1B**-launched command and service modules remaining from the **Apollo** lunar programme.

The station comprised an orbital workshop (OWS) and attached instrument unit, adapted from a Saturn V (S-IVB) third stage; an **airlock** module (AM); a multiple docking adapter (MDA), which allowed up to two Apollo spacecraft to **dock**; and the Apollo telescope mount (ATM), designed for solar observations in particular. It had an oxygen/nitrogen atmosphere at a pressure of about 34 kPa (5 psi) and a total habitable volume of about 360 m^3. Two **solar panel**s with a combined area of 11.8 m^2

and capable of producing nearly 12 kW were attached to the OWS and four more of 111.5 m² to the ATM. Launch vibrations caused the dislocation of a combined **meteoroid**/thermal shield which tore away one of the OWS arrays and restricted the deployment of the other. The first Skylab crew were, however, able to restore the station to an **operational** condition. Despite attempts to raise its orbit, Skylab re-entered the Earth's atmosphere on 11 July 1979 and spread its debris over a thinly populated area of Western Australia.
[See also **Salyut, Mir.**]

Figure S5 The American **Skylab** space station in Earth orbit. Note the sunshield, deployed by the first crew, and the single large **solar array** panel. The other panel was torn off during the launch. [NASA]

SLANT RANGE

The distance between a **launch vehicle** or a **spacecraft** in **orbit** and the ground station tracking it. Used particularly for launch vehicles, the term recognises that the **range** is being measured in a direct line to the vehicle without regard to its position relative to the ground (e.g. it may be only 3 km east of the **launch pad**, but its **altitude** may be 4 km: this makes its range 5 km, along the 'slanted' hypotenuse of the triangle thus formed).
[See also **downrange.**]

SLEEP RESTRAINT

See **restraint**.

SLIDE-WIRE

An escape device attached to a launch vehicle **service structure**, comprising a 'cage' or 'basket' that can slide down an inclined wire, to take **astronaut**s away from the **launch pad** in an emergency.

SLING-SHOT (GRAVITATIONAL)

See **gravitational boost**.

SLOSHING

A rhythmic lateral motion of a **liquid propellant**, particularly in the tanks of a **launch vehicle**, which can be caused by the vehicle's motion and can lead to instability and perturbations in the flight **trajectory**. [See also **anti-slosh baffle, pogo**.]

SLOW-SCAN TV

A television system which features a reduced number of picture 'frames' per second (i.e. less than the standard 25). Since the amount of picture information is reduced, this system requires a lower transmission **bandwidth**, which can reduce transmission charges (via satellite or otherwise). In fact a typical **bit rate** for slow-scan TV is 64 kbits s^{-1} compared with 68 Mbits s^{-1} for 625-line colour TV.

The main drawback of the system is that any significant motion produces a blur, due to the low 'refresh rate' of the picture. The system is, however, useful for applications such as videoconferencing where subject movement is kept to a minimum. The typical bit rate for videoconferencing is about 2 Mbits s^{-1}.
[See also **videoconference**.]

SLOW-WAVE STRUCTURE

That component of a **travelling wave tube** (TWT) which reduces the velocity of an **RF wave** carried along it so that it travels slightly slower than the **electron beam** [see figure T5]. This allows an interaction between the RF wave and the electron beam resulting in an energy transfer from the beam to the wave—the RF amplification function.

A slow-wave structure may take the form of a helix or of coupled cavities. The helix, made of copper or of tungsten or molybdenum wire, is supported by three or four ceramic rods to isolate the RF fields on the helix from the metallic walls of the surrounding vacuum envelope. The coupled-cavity type comprises accurately sized and shaped RF cavity sections, usually of copper, brazed together to form a structure of many cavities 'in cascade'.

SLUSH HYDROGEN

A potential **cryogenic propellant**: a mixture of solid and liquid hydrogen which is even colder than **liquid hydrogen**.

SM

See **structure model**.

SMALL SELF-CONTAINED PAYLOAD (SSCP)

The formal name for a **getaway special** (GAS) payload.

SMATV

An acronym for satellite master antenna television; television distributed to cable **head-end**s by satellite.
[See also **CATV**, **cable TV**.]

SNOOPY CAP

A colloquial term for the 'communications carrier assembly', a close-fitting head cap worn by **astronaut**s, typically inside a **spacesuit** helmet, which includes headphones and a microphone. Named after the flying helmet worn by the cartoon character 'Snoopy'.

SOFT DOCK

See **dock**.

SOFT FAIL

Jargon: the failure of a component, device or **subsystem** which does not seriously degrade the performance or **capacity** of a **system**.
[See also **degradation**, **graceful degradation**.]

SOFTWARE

Any form of coded information used in the operation of a computer system; a computer program; the non-physical part of a computer-based system (cf **hardware**).

SOLAR ABSORBER

See **surface coatings**.

SOLAR ARRAY

An area of **solar cell**s designed to provide electrical power to a spacecraft. Individual cells are connected in series to provide the voltage required for a particular **power** subsystem design, and the series strings are connected in parallel to form an 'array section'. This construction helps to ensure that single cell failures do not 'short out' whole sections of the array. Due to the large area of cells required for most applications, the majority of arrays are stored folded against the spacecraft for launch and **deploy**ed in space. Mechanically speaking, arrays are of two different types—rigid or flexible.

Currently most common is the rigid-panel type. In the **three-axis-stabilised** spacecraft design, a hinged array of flat panels is connected to the spacecraft by a lightweight yoke structure attached to the solar array drive mechanism. Most three-axis satellites have two panels mounted perpendicular to their north and south faces. In order to remain pointing at the Sun, they rotate once per day about the satellite's north–south axis (the **pitch axis**).

The arrays on the lower-power **spin-stabilised** spacecraft form the outer skin of the spacecraft body; spinners with higher power capabilities have an additional cylindrical 'skirt', mounted over the prime array, which is deployed in a telescopic fashion.

Figure S6 NASA **solar array** experiment (SAE) flown on STS 41-D. The white circles are targets for a stereo TV system which recorded the motion of the array when the **orbiter**'s attitude control thrusters were fired. [NASA]

Flexible arrays or 'array blankets', for use on three-axis-stabilised spacecraft, are typically larger and of lower density. They can be stored in a box which is swung out from the side of the spacecraft on a hinged arm. An extendable truss structure, for example an '**Astromast**', then pulls the array blanket from its box. The NASA solar array experiment (SAE) [figure S6] flown on the STS 41-D mission was the largest example at the time: it was 32 m long, 4 m wide and comprised 84 panels together producing some 12.5 kW. The array was folded, concertina fashion, into a stack only 19 cm thick, 0.6% of its extended length.

The power-generation capability of solar arrays is dependent upon certain astronomical mechanisms. Due to a slight eccentricity of the Earth's orbit, the Sun—Earth distance is greatest at the summer **solstice** and least at the winter solstice. This produces a seasonal variation in solar flux density. More important, though, is the **declination** of the Sun with respect to the solar array surface which, for the usual orientation of an array surface parallel to the Earth's axis, is 0° at the **equinox**es and 23.5° at the solstices. Maximum power can only be derived if the array surfaces are perpendicular to the solar vector.

[See also **BAPTA**, **fuel cell**, **radioisotope thermoelectric generator**, **spacecraft axes**, **degradation**.]

SOLAR ARRAY DRIVE MECHANISM (SADM)

See **BAPTA**.

SOLAR CELL

A device for the conversion of solar energy to electrical energy.

Also known as a photovoltaic cell, since it operates in accordance with the photovoltaic effect, whereby **electromagnetic radiation,** incident on a thin film of n-type semiconductor material on the surface of a p-type semiconductor, produces a potential difference between the two [figure S7].

Silicon is the predominant semiconductor used in solar cells. The addition of phosphorus (a process known as 'doping') results in an n-type semiconductor; silicon doped with boron is a p-type material ('n' stands for negative and 'p' for positive). Solar photons liberate excess **electron**s from the upper layer and their movement to the electron-deficient lower layer constitutes a current.

Silicon cells are typically made from ingots of pure silicon doped with boron. The ingot is cut into discs and phosphorus is diffused into the top surface to make a semiconductor (p/n) junction across the surface of the cell. For spacecraft applications the discs are then cut into rectilinear shapes to increase the packing density of the resultant

solar array. Other semiconductors used for solar cells include gallium arsenide (GaAs) and indium phosphide.

The first practical solar cell was developed at Bell Telephone Laboratories (USA) in 1954 and the first satellite to make use of it was Vanguard 1, launched by the USA in March 1958. [See also **coverslip**.]

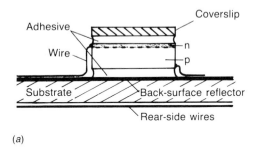

(a)

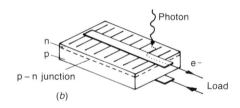

(b)

Figure S7 The structure of a typical **solar cell**.

SOLAR CONSTANT

A measure of the incident solar energy at the extremity of the Earth's atmosphere; $S = 1.37$ kW m^{-2} (± 0.02 kW) at 1 **astronomical unit** (AU).

SOLAR FLARE

A sudden eruption on the surface of the Sun resulting in an increased flux of high-energy particles—see **solar wind**. The flares cause radio and magnetic disturbances on the Earth and increase the activity of the **magnetosphere** and the **ionosphere**. As a result, they threaten the crew of manned spacecraft with potentially damaging ionising radiation. The structural shells of spacecraft, and especially **space station**s where exposure may be more prolonged, must be sufficiently thick to

reduce radiation to tolerable levels.
[See also **Van Allen belts**.]

SOLAR GENERATOR

An alternative name for **solar array**.

SOLAR PANEL

An individual element of a flat-panel **solar array**.

SOLAR REFLECTOR

See **surface coatings**.

SOLAR SAILING

The use of the **solar wind** in controlling the position and orientation of spacecraft in orbit or on interplanetary trajectories.

The **solar array**s of a **three-axis-stabilised** satellite, mounted perpendicular to its north and south faces, rotate once per day about the satellite's north–south (**pitch**) axis to remain pointing at the Sun. This allows them to be used as an **actuator** for the **attitude and orbital control system**, providing adjustments in **roll** and **yaw**. If one of the array panels is driven to a position parallel to a line drawn from the Sun, the radiation pressure on the other will exert a moment on the spacecraft and it will begin to rotate about an axis orthogonal to the N–S axis; swopping the orientation of the arrays would rotate it in the opposite direction. Further, by orientating the arrays in the manner of an aircraft propeller, the spacecraft can be rotated about the third orthogonal axis, producing a '**windmill torque**'. In practice, flaps attached to the arrays allow array angles to be minimised, which keeps the power loss due to array mispointing to about 1%.

Solar sailing has also been suggested as a method of interplanetary propulsion, mainly in the inner Solar System where the solar wind has an appreciable magnitude. No doubt the techniques used would be analogous to those used in terrestrial sailing.
[See also **spacecraft axes**.]

SOLAR SPECTRUM

The distribution of **electromagnetic radiation** emitted by the Sun. Figure S8 shows how the Sun's output is spread over the frequency spectrum: approximately 7% is at ultraviolet **wavelength**s (<0.38 μm); 45.5% is in the visible ($0.38 - 0.76$ μm); and 47.5% is in the infrared and beyond (>0.76 μm). The 'spectral energy density' of the Sun is

important in the design of spacecraft **surface coatings**, amongst other things.
[See also **thermal control subsystem**.]

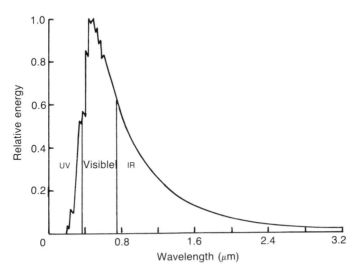

Figure S8 A graphical representation of the **solar spectrum**.

SOLAR WIND

A continuous stream of atomic and other particles ejected from the Sun, otherwise known as solar radiation pressure; one of the mechanisms that results in the perturbation of a spacecraft's **orbit**.

Depending on solar activity, the velocity of the particles, or **plasma**, varies between about 300 and 500 km s^{-1}, the force imparted being sufficient to affect spacecraft with large surface areas, for instance those with large **solar array** panels. The solar wind is thus one of the **perturbations** affecting the **station keeping** of satellites in **geostationary orbit**. It can, however, be used to effect for spacecraft **attitude control**: see **solar sailing**.

It is, incidentally, the solar wind that causes the tails of comets to orient themselves in a direction away from the Sun wherever they might be in their orbit about the Sun. It also causes an elongation of a planet's **magnetosphere** [see figure V1].
[See also **solar flare, equilibrium point, triaxiality, luni–solar gravity**.]

SOLID PROPELLANT

The combustible substance which provides **thrust** in a **solid rocket motor**. There are two main types of solid propellant: the homogeneous or double-base propellants (nitrocellulose plasticised with nitroglycerine plus stabilising products), which are limited in power and not widely used in space applications; and the heterogeneous or composite type.

Composite propellants consist of a mixture of **fuel** and **oxidiser**. An oxidiser in common use is crystalline ammonium perchlorate (NH_4ClO_4; abbreviated to AP). It is mixed with an organic fuel like polyurethane or **polybutadiene**, which also binds the two components together. The inclusion of a plastic 'binder' produces a 'rubbery' material, making the propellant relatively easy to handle. Performance is enhanced by the addition of finely ground (10 μm) metal particles of aluminium, for example, which increase the heat of the reaction due to the formation of metal oxides, and increase the overall density of the propellant. As an example, the propellant used in the Space Shuttle **solid rocket booster**s comprises 14% polybutadiene acrylic acid acrylonitrile (binder/fuel), 16% aluminium powder (fuel), 69.93% NH_4ClO_4 (oxidiser) and 0.07% iron oxide powder (catalyst).

Solids are generally easy to handle and store, and do not corrode their containers as some liquid propellants do. The solid motor is simpler, but operationally less flexible [see **rocket motor**].

[See also **propellant grain, propellant, liquid propellant**.]

SOLID ROCKET BOOSTER (SRB)

(i) A propulsive component of a **launch vehicle** which uses **solid propellant**. An SRB added to a **core vehicle** to augment its **thrust** is otherwise known as a 'strap-on' booster.

(ii) Specifically, one of two boosters strapped to the **external tank** of the American **Space Shuttle** launch vehicle. The SRBs burn with the **Space Shuttle main engine**s for about two minutes into the **flight** and are then jettisoned, parachuted into the ocean and recovered for refurbishment and re-use. Each booster is 38.5 m long and 3.7 m in diameter, contains about 500 tonnes of **propellant** [see **solid propellant**], and provides a thrust of approximately 12 million newtons (*in vacuo*) with a **specific impulse** of about 267 s. The **nozzle** can be gimballed for steering ($\pm 8°$ from the booster axis).

These boosters were the first solid rocket boosters designed to be used with a manned vehicle; the failure of an 'O-ring' seal between two SRB segments was identified as the cause of the so-called '**Challenger disaster**' (mission STS 51-L) in January 1986.

(iii) Another name for a **solid rocket motor (SRM)**.
[See also **rocket motor**.]

SOLID ROCKET MOTOR (SRM)

See **rocket motor**.

SOLID STATE POWER AMPLIFIER (SSPA)

A device, using semiconductor materials or components, for the amplification of **radio-frequency** (RF) signals. Utilised in spacecraft **communications payload**s as an alternative to the **travelling wave tube amplifier** (TWTA). Developments in SSPAs have shown them to improve significantly the performance and reliability of the **transmit chain** of a **transponder**. However, since the solid state medium is a much poorer conductor of heat than the copper (say) in TWTs, SSPAs require large radiators for higher output powers. The excessive weight limits SSPAs to relatively low-power applications, while TWTAs provide the higher power amplification.

SOLSTICE

Either of two annual occasions when the Sun is overhead at the tropic of Cancer (summer solstice, 21 June) or the tropic of Capricorn (winter solstice, 21 December); the longest and the shortest days of the year, respectively.
[See also **equinox, eclipse, ecliptic, solar array**.]

SONDE

See **sounding rocket**.

SOUND-IN-SYNCS (SIS)

A type of TV audio transmission in which the sound is coded into the line-synchronisation pulse of the **television** waveform.

SOUND SUPPRESSION SYSTEM

A **launch pad** facility which protects a **launch vehicle** and its **payload**s from acoustic energy produced by a **rocket motor** at and following **ignition** and reflected from the surface of the **launch platform**. Typical systems comprise volumes of standing water and water sprays, the latter of which also serve to constrain the temperature of the structure and avoid damage. It is the evaporation of this water which produces the large clouds of steam often visible at rocket launches.

SOUNDING ROCKET

A **rocket** capable of carrying a small **payload** on a **sub-orbital trajectory**, used mainly for the investigation of the Earth's upper atmosphere. Rocket- or balloon-borne instruments are sometimes called 'sondes' (named by analogy with measurements of water depth: 'soundings').

SOYUZ

A series of Soviet **manned spacecraft** developed in the 1960s and used in improved forms into the 1990s [figure S9]. It comprises three **module**s: a spherical orbital compartment at the forward end, containing **life support** equipment and used for scientific experiments, etc; a **re-entry** module, containing the cosmonaut's couches and flight control equipment; and an equipment module (similar in function to the **Apollo** spacecraft's service module) with two **solar array** panels mounted on it.

Figure S9 A Soviet **Soyuz** spacecraft in Earth orbit on the **Apollo—Soyuz test project** mission in July 1975. Part of the docking mechanism is visible at the front of the spacecraft (right). [NASA]

The first manned Soyuz mission was launched 23 April 1967. Thereafter, this first variant of the spacecraft was used increasingly for **docking** experiments and missions which placed multiple crews in orbit, and later to deliver crews to the Soviet **space station**, **Salyut**. It originally carried a crew of three, but this was reduced to two following the Soyuz 11 mission in April 1971, when the crew, who were not wearing **spacesuit**s, died on return from orbit when the spacecraft depressurised. The second Soyuz variant, Soyuz-T, first launched 27

November 1980, was capable of carrying a crew of three (with spacesuits); it was later updated further to the Soyuz TM (first flight May 1986).

SPACE

(i) The three-dimensional expanse in which all material objects are located.

(ii) The region beyond the limits of the Earth's atmosphere which is the operating environment of a **spacecraft** ('near-Earth space', 'interplanetary space', **deep space**).
[See also **space—time, degree of freedom, frame of reference.**]

SPACE ASTRONOMY

Astronomy conducted from space, using observation equipment in Earth **orbit**, or in orbit about or on the surface of a **planetary body**, such as the Moon, which has no appreciable atmosphere. The observation platform may be a dedicated **astronomical satellite**, or a **module** or **payload** of a **space station**. The main advantage of space astronomy is the position of the **observatory** above the Earth's atmosphere, which perturbs and attenuates the **image**.
[See also **atmospheric attenuation.**]

SPACE CAPSULE

See **capsule**.

SPACE DEBRIS

See **orbital debris**.

SPACE PLATFORM

A general term for a **space station** or similar (usually large) **spacecraft** in a permanent orbit. See **platform**.

SPACE-QUALIFIED

An attribute of any component or device considered able to perform its function in the space environment. The term is understood on two different levels: (i) a space-qualified component is one which has flown (and performed to specification) in space; and (ii) it is one which has been tested on Earth in a simulated space environment to a point where it is considered capable of performing in space.

SPACE SCIENCE

That branch of science concerned with the features and processes of the

space environment. It includes 'space physics', '**space astronomy**', etc, and can be extended to disciplines with applications in space (e.g. 'space medicine' and **microgravity** research).

SPACE SEGMENT

The part of a **satellite**-based communications system or any **spacecraft** communications link which resides in space. The space segment may comprise more than one satellite. The term may also be used for any space-based hardware, in order to differentiate it from the complementary hardware on the ground, the **earth segment**.

SPACE SHUTTLE

In general, a reusable **spacecraft** or **launch vehicle** designed to carry personnel and/or **payload**s into orbit, return for refurbishment and repeat the exercise many times throughout its **operational** lifetime. In particular, the American Space Shuttle, otherwise known as the **space transportation system** (STS), which comprises an **orbiter**, **external tank** (ET) and two **solid rocket boosters** (SRBs) [figure S10]. In the case of the STS the orbiter alone is considered a 'spacecraft'; the orbiter with its ET and SRBs is a 'launch vehicle'.

Figure S10 Launch of the American **Space Shuttle**. [NASA]

Although many design concepts for a reusable space transportation system appeared in the 1960s, the present concept dates from the 1970s: the contracts to design and develop the Shuttle orbiter were announced in July 1972, and those for the ET and SRBs in 1973 and 1974 respectively. Approach and landing tests of the orbiter **Enterprise** were made in the late 1970s and the first spaceflight began on 12 April 1981 (STS-1: **Columbia**).

[See also **Atlantis, Challenger, Discovery, Endeavour, Buran, payload bay, flight deck, Space Shuttle main engine (SSME), auxiliary power unit (APU), orbital manoeuvring system (OMS), remote manipulator system (RMS), thermal protection system (TPS), solid propellant, specific impulse.**]

SPACE SHUTTLE ABORT ALTERNATIVES

See **abort**.

SPACE SHUTTLE MAIN ENGINE (SSME)

One of three **rocket engine**s used, along with two **solid rocket booster**s, to launch the American **Space Shuttle** [see frontispiece]. The engines are mounted aft of the **orbiter**'s **payload bay** in a triangular configuration: with the orbiter horizontal and viewed from the rear the uppermost engine is designated 'No 1', the left-hand engine 'No 2' and the right-hand engine 'No 3'. The SSME uses the **cryogenic propellant**s **liquid oxygen** and **liquid hydrogen** (in a 6:1 LOX/LH$_2$ mixture ratio) and develops a nominal (100%) **thrust** of 2.1 million newtons (470 000 lb) *in vacuo* (1670 000 N (375 000 lb) at sea level). The engines are typically operated at the 104% thrust level in the latter stages of a flight [see **thrust profile**].

The engines are individually gimballed to steer the Shuttle on its ascent and regeneratively cooled by the liquid hydrogen [see **gimbal, regenerative cooling**].

[See also **combustion chamber, nozzle, ignition system.**]

SPACE STATION

(i) A **manned spacecraft** designed to remain in **orbit** and provide permanent, or semi-permanent, accommodation and associated **life support system**s for a **crew** who may inhabit the spacecraft for long periods [see **Skylab, Salyut, Mir**].

(ii) In the late 1980s, the term generally used to refer to the **NASA** International Space Station, called '*Freedom*', a collaborative project with Europe, Canada and Japan [figure S11].

(iii) In the terminology of official regulatory bodies (e.g. the **ITU**), a 'space station' is any **spacecraft** 'stationed in **space**' (i.e. a **satellite**), as opposed to an **earth station**).

SPACE TECHNOLOGY

The application of the physical and, to a lesser extent, biological sciences to the investigation, exploration and industrial/commercial utilisation of **space**; the methods, theory and practices governing such application.

SPACE TELESCOPE

Generally any telescope in Earth orbit, but usually refers to the Hubble Space Telescope (HST), named after Edwin Powell Hubble (1889–1953),

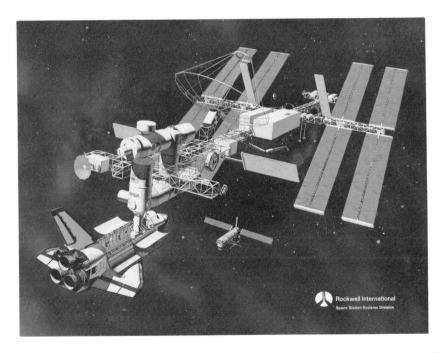

Figure S11 An artist's impression of a design concept for the US/International **Space Station**. The large flat panels are **solar array**s, extended by devices similar to the **astromast** depicted in figure A10; the smaller panels are **radiator**s. The station comprises a number of **habitable module**s and a truss structure supporting, amongst other things, a large **unfurlable antenna** (top). Note also the Shuttle docked at lower left, its radiator panels standing proud of the payload bay doors, and the free-flying space **platform** below the Station. [Rockwell International]

the American astronomer [figure S12]. Designed to be launched and retrieved from **low Earth orbit** by the **Space Shuttle**, the HST is an optical telescope of the Cassegrain type, with a 2.4 m diameter mirror [figure S13].
[See also **Cassegrain reflector**.]

Figure S12 An artist's impression of the Hubble **Space Telescope** in Earth orbit. Note the twin **solar arrays**, the communications **antenna**s and the (open) protective cover at the end of the tube. [Lockheed]

SPACE–TIME

In physics, the four-dimensional continuum having three spatial coordinates and one time coordinate which completely specify an event. Also called 'space–time continuum'. Anything existing in both space and time, or concerned with space–time, is referred to as 'spatiotemporal'.
[See also **space**.]

SPACE TRACKING AND DATA NETWORK (STDN)

A worldwide network of **tracking** stations for the provision of **communications** (i.e. **telemetry** and voice links) with **spacecraft** in Earth **orbit**. The network is operated by **NASA**'s Goddard Space flight

Center (GSFC) in Greenbelt, Maryland and supplemented by the tracking and data relay satellite system (**TDRSS**).
[See also **deep space network (DSN)**.]

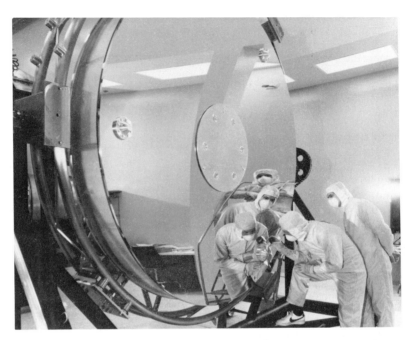

Figure S13 The mirror for the Hubble **Space Telescope**. It is 2.4 m in diameter, weighs 820 kg and is coated with aluminium and magnesium fluoride. [NASA]

SPACE TRANSPORTATION SYSTEM (STS)

The formal name for the American **Space Shuttle**. Following the first four 'test flights', designated STS-1 to STS-4, the STS mission designation took the form (for example) STS 41-D. The first figure represented the final digit of the fiscal year for which the launch was planned (1984 in this example); the second figure represented the **launch site**, where 1 was **Kennedy Space Center** (KSC) and 2 was **Vandenberg Air Force Base** (VAFB)†; and the letter represented a particular flight in that fiscal year, 'D' being nominally the fourth flight. This system allowed payloads to be allocated to a particular **orbiter** as opposed to a specific position in the flight order, which meant that if a **payload** or orbiter suffered a delay, another mission which *was* ready

†Note: in fact there were no Space Shuttle launches from Vandenberg in the 1980s: after the Challenger failure the VAFB Shuttle launch complex was 'mothballed' for an indefinite period.

could be launched instead. This system of flight designation was retained until the **Challenger** failure in January 1986, whereupon it reverted to the simple STS-26, STS-27, etc, designation, due largely to the significant reduction in the flight rate.

SPACE TUG

See **orbital manoeuvring vehicle**.

SPACE VEHICLE

See **spacecraft**.

SPACECRAFT

Any self-contained vehicle designed for **spaceflight**; alternatively known as a 'space vehicle' [but see **vehicle**]. The term functions as both singular and plural.

To qualify as a 'spacecraft', a vehicle typically has, as a minimum, a **propulsion subsystem**, a **power supply** and a **payload** [see **manned spacecraft, unmanned spacecraft, man-tended spacecraft**]. An exception is the **passive communications satellite**. Although **launch vehicle**s meet these requirements, they are not designed for spaceflight and do not qualify as spacecraft.

Any vehicle which is also capable of 'flight' within the atmosphere is termed an '**aerospace vehicle**'.
[See also **manned manoeuvring unit (MMU), attitude and orbital control, thermal control subsystem, telemetry tracking and command (TT&C), life support system**.]

SPACECRAFT AXES

The three orthogonal axes of rotation: roll, pitch and yaw.

If a vehicle has a recognisable longitudinal axis or a specified 'forward direction of flight', the axes are analogous to those of an aircraft: the roll axis is the longitudinal axis; the pitch axis is the one in the plane of the wings; and the yaw axis is the 'vertical' axis, orthogonal to both the roll and pitch axes. The axes are mutually perpendicular, with an 'origin' at the vehicle's **centre of mass**.

For a winged spacecraft, like a **Space Shuttle**, the similarity with an aircraft is obvious. For **expendable launch vehicle**s the roll axis is the axis which is vertical at launch and the other axes are more or less arbitrarily assigned since the vehicle is rotating about the roll axis [see **roll programme**, but see also **pitch-over**].

The axes of a cylindrical spacecraft (e.g. **Apollo, Soyuz**, etc) are similar to those of an ELV at launch, but, once in orbit, assume the axis definition of an aircraft (i.e. defined relative to the pilot's seat).

The axes of a **satellite** mirror those of an aircraft 'flying along the orbital arc': the roll axis is aligned with the direction of travel; the yaw axis passes through the **sub-satellite point**; and the pitch axis is orthogonal to the other two [figure S14]. For a satellite in an **equatorial orbit**, the pitch axis is aligned approximately with the Earth's spin axis. The pitch axis is also the spin axis for the **spin-stabilised** satellite.

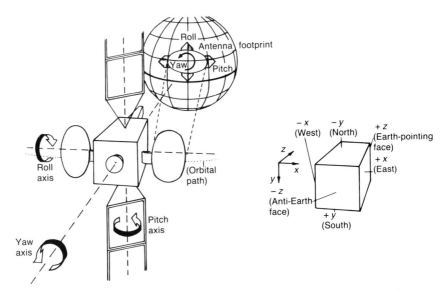

Figure S14 The axes and the orientation of a **three-axis-stabilised** satellite in **geostationary orbit**—see **spacecraft axes**.

In the compilation of engineering drawings the three orthogonal axes are often labelled in cartesian fashion : x = roll, y = pitch, z = yaw. For the **three-axis-stabilised** satellite, the x axis and y axis are otherwise known as the east—west and north—south axes, respectively; the z axis points towards the sub-satellite point.

[See also **degree of freedom, barbecue roll, dutch roll, actuator**.]

SPACECRAFT CABIN

See **cabin**.

SPACECRAFT INTEGRATION

The assembly of the components of a **spacecraft** prior to final ground testing and **launch**. For a **satellite** it typically involves the joining of the **payload module, service module, antennas** and **solar arrays**.

[See also **payload integration**.]

SPACEFLIGHT

The act of travelling through space; a particular journey or mission in space. See **spacecraft**.

SPATIAL FREQUENCY REUSE

See **frequency reuse**.

SPACELAB

A manned laboratory developed by the European Space Agency (**ESA**) for flights aboard the American **Space Shuttle** [figure S15]. A particular Spacelab **payload** can include various sizes of 'pressurised **module**' and a number of '**pallet** segments' dependent upon the specific **mission**. If used, the pressurised module is connected to the **cabin** of

Figure S15 The preparation of **Spacelab** 1 for a **Space Shuttle** mission. Note **habitable module** at top and **pallet** carrying scientific instruments below. [NASA]

the **orbiter** by a tunnel, which can be fitted with an **airlock** for **extra-vehicular activity** (EVA) if required. Materials and equipment designed to be exposed to space are mounted on the pallets. If Spacelab is configured as a pallets-only mission, the power, experiment control and **data**-handling equipment is protected within a pressurised and temperature-controlled housing called an 'igloo'.

SPACELAB PALLET

See **pallet**.

SPACEPORT

A facility engaged in the launching of **spacecraft** and other connected activities [see **Baikonur Cosmodrome, Guiana Space Centre, Jiuquan, Kennedy Space Center (KSC), Northern Cosmodrome, San Marco platform, Tanegashima, Vandenberg Air Force Base (VAFB), Xichang**].
[See also **launch complex, launch vehicle**.]

SPACEPROBE

See **probe**.

SPACESUIT

A pressurised suit worn by **astronaut**s as protection against the space environment (vacuum, thermal extremes and radiation). Also called a 'pressure suit' or 'extra-vehicular pressure garment'. **NASA** spacesuits are commonly known as extravehicular mobility units (EMUs) [see figure S16]. Spacesuits for use inside the **cabin** (e.g. in case of cabin **decompression**), which are pressure tight but have fewer protective layers, are known as 'intra-vehicular pressure garments'.

The EMUs used on the American **Space Shuttle**, to take a modern example, comprise a zip-on, one-piece undergarment and a two-piece pressurised spacesuit. The former is known as a liquid cooling and ventilation garment (LCVG). It includes urine collection facilities, a 'drink bag' and a **'snoopy cap'**, with headphones and a microphone, which fits over the head and chin. The spacesuit itself consists of upper and lower torso sections, joined at the waist in a hard ring, and a helmet. A visor assembly, which offers further protection against **micrometeoroid**s and ultraviolet and infrared radiation, fits over the helmet. The total weight is about 30 kg. The suit has several bonded layers: a polyurethane-on-nylon pressure bladder, several layers of Kevlar, and an outer abrasion-resistant layer of Kevlar, Teflon and

Dacron. The hard upper torso has an **aluminium** shell. The torso sections are made in a variety of sizes for use by different crewmembers, unlike previous American spacesuits which were tailored for individual astronauts.

[See also **portable life support system, pre-breathe, Kevlar composite.**]

Figure S16 Astronaut Bruce McCandless flys the **manned manoeuvring unit** (MMU) during the STS 41-B **Space Shuttle** mission [see also figure M1]. He is wearing an 'extravehicular mobility unit' (EMU)—see **spacesuit**. Note the black light-pipes which curl in front of the helmet—they provide an optical signal when the MMU thrusters are firing. [NASA]

SPACEWALK

See **extra-vehicular activity (EVA).**

SPATIOTEMPORAL

See **space–time.**

SPECIFIC ENERGY

A figure of merit for a power source (e.g. a **battery**), measured in watt hours per kilogram (W h kg^{-1}). Since this is a measure of the usable **energy** stored per unit mass, it is sometimes called the 'usable energy density'.

[See also **specific power**.]

SPECIFIC IMPULSE

(i) The **impulse** per unit mass of **propellant** consumed; a measure of the efficiency and performance of a **rocket engine** or **rocket motor**.

Specific impulse is given by the expression

$$I_{sp} = I/M$$

where impulse, I, is in N s, and mass, M, is in kg. I_{sp} is therefore measured in N s kg^{-1}, which can be reduced to m s^{-1}.

(ii) An alternative, but equivalent, definition is: the ratio of **thrust** (N) to rate of propellant consumption (kg s^{-1}), which gives the same units for I_{sp} (m s^{-1}).

Historical note: originally I_{sp} was defined as impulse per unit weight of propellant consumed, I/Mg, with I in newton seconds and Mg in newtons. This gave the second (s) as the unit of specific impulse. In terms of definition (ii), when thrust was measured in pounds force and rate of consumption in pounds per second, the units (lb (lb s^{-1})$^{-1}$) also gave the second as the unit for I_{sp}. This was a convenient concept, since the specific impulse could be said to be the time (in seconds) over which the combustion of 1 kg of propellant produced a thrust of 1 kg: the longer the time, the less fuel used to produce an equivalent thrust and the more efficient the engine; the longer the time, the greater the specific impulse.

Specific impulse is still often quoted in seconds. The different types of propellant used in the American **Space Shuttle** system are indicative of contemporary achievable I_{sp}: the **solid rocket boosters** develop on average 250 s dependent upon the altitude; the **Space Shuttle main engines**, which use LOX and LH$_2$, are rated at about 363 s at sea level and 453 s *in vacuo*; the orbital manoeuvring system (OMS) and reaction control system (RCS), which use the common **bipropellant** combination of **nitrogen tetroxide** and **MMH**, give around 315 s and 270 s respectively. In comparison, the **Saturn V**'s second and third stages (burning LOX and LH$_2$) had an I_{sp} of 425 s, somewhat less than the more advanced SSMEs, and its first-stage engines, burning LOX and **kerosene**, produced an I_{sp} of 265 s.

I_{sp} quoted in seconds can be converted to m s^{-1} by multiplying by 'g', the acceleration due to **gravity** (9.81 m s^{-2}).
[See also **exhaust velocity**.]

SPECIFIC POWER

A quantitative measure of the mass efficiency of a **power** subsystem or device (units: W kg^{-1}). For example, for a **solar array** it is expressed as the ratio of array power output to array mass.
[See also **specific energy**.]

SPEED OF LIGHT

See **light (velocity of)**.

SPEEDBRAKE

A device on the tail fin of the American Space Shuttle **orbiter**, used as a rudder for the atmospheric part of the descent from orbit and split vertically into two parts on landing to increase **drag** and assist deceleration.

SPELDA

A French acronym for Structure Porteuse Externe pour Lancements Doubles Ariane (external support structure for Ariane dual launches).

A container mounted on the third stage of the **Ariane** 4 launch vehicle, below the **shroud**, which allows two independent spacecraft to be carried and launched by a single vehicle. One spacecraft is contained within the SPELDA, another is mounted on top. It differs from the **SYLDA** in that the SPELDA is part of the external structure or **fairing** of the vehicle, and not simply a container mounted inside the shroud. To accommodate triple **payload**s a SYLDA can be mounted on top of a SPELDA.
[See also **SPELTRA**.]

SPELTRA

A French acronym for Structure Porteuse Externe pour Lancements TRiples Ariane (external support structure for Ariane triple launches).

A container designed to be mounted on the third stage of the **Ariane** 5 launch vehicle, below the **shroud**, which allows three independent spacecraft to be carried and launched by a single vehicle.
[See also **SPELDA, SYLDA**.]

SPF

See **single-point failure**.

SPILLOVER

The part of an **antenna** feed's radiation pattern not intercepted by the antenna reflector or **subreflector**.
[See also **overspill**.]

SPIN-STABILISED (SPACECRAFT)

A spacecraft which maintains its stability by rotating about its longitudinal axis [see **spacecraft axes**] [see figures S17 and S18].
[See also **three-axis-stabilised**, **gyroscopic stiffness**, **de-spun platform**, **'spin-up'**.]

Figure S17 An artist's impression of the Intelsat VI **spin-stabilised** communications satellite. Note the **deployable antennas**—compare with figure D2. [Hughes Aircraft Company]

SPIN TABLE

A device mounted on a **launch vehicle** which 'spins up' its spacecraft

payload prior to release, in order to provide spin stabilisation in the **transfer orbit** (used for example on the American **Space Shuttle**, **Delta** and **Titan** launch vehicles). By contrast, the **Ariane** vehicle provides spin by spinning the whole third **stage**. (Note: some satellites are **three-axis-stabilised** in the transfer orbit).
[See also **spin-up**, **spin-stabilised**.]

Figure S18 The SBS **spin-stabilised** communications satellite. [Hughes Aircraft Company]

SPIN-UP

(i) The rotation of a **spin-stabilised** spacecraft prior to its release from the **Space Shuttle** payload bay or from other **launch vehicles** using a **spin table**.

(ii) The increase in rotation rate of a spin-stabilised spacecraft (to its operational spin rate) at the beginning of its life in orbit.

(iii) The undesirable rotation of a spin-stabilised spacecraft's **de-spun platform** resulting in a loss of 'earth-lock'.

SPINNER

A colloquial term for a **spin-stabilised** spacecraft.

SPINNING SOLID UPPER STAGE

See **payload assist module (PAM)**.

SPLASHDOWN

The moment at which a **spacecraft** returning to Earth lands in a body of water which, for most American spacecraft, is the Pacific Ocean. Early US **manned spacecraft** (i.e. **Mercury**, **Gemini** and **Apollo**) returned to a water landing; the equivalent term for a vehicle returning to a body of land is '**touchdown**'.

SPLASHPLATE

A type of antenna **feed** system incorporating a generally flat **subreflector**, rather than one with a precise mathematical profile (e.g. hyperbolic). Also called a 'ring-focus' feed, because the phase centre of the wave takes the form of a ring between the splashplate and the feed device.

SPOT BEAM

A **beam** formed by a **satellite** communications **antenna** which provides coverage of a narrowly defined area: a small country or part of a larger one. Spot beams concentrate the satellite's **radiated power** in relatively small areas, which increases the **power flux density** at the ground. This may be used to decrease the necessary size of receiving antennas or, alternatively, to give higher traffic **capacity** and/or wide **bandwidth** services.

[See also **beamwidth**, **global beam**, **hemispherical beam**, **zone beam**, **multiple beam**, **traffic**.]

SPREAD SPECTRUM

In **telecommunications**, a transmission technique which spreads the **data** over a **bandwidth** many times that required for the original data. Distributing the power reduces the **amplitude** of the **signal** to such an extent that it disappears into the **noise**, allowing the residual capacity of an allocated **transponder** to be used as well as dedicated transponders. This technique is useful for low data rates and small receive antennas. It is used primarily for military communications since it offers a degree of protection against jamming.

SPREADING LOSS

A loss in power density experienced by a telecommunications signal as it radiates from its point of origin. If **power** is radiated isotropically from a point source, forming a spherical wavefront, the power density on the wavefront diminishes progressively with distance (obeying an inverse square law). In calculating the loss of signal power in a practical telecommunications system, this natural tendency is incorporated with **frequency** to give an expression for **free space loss**. [See also **isotropic**, **equivalent isotropic radiated power (EIRP)**.]

SPURIOUS SIGNAL

A **signal** in a communications system which is not intended; one which has arisen by spurious means, not always understood.

SPUTNIK

The name given to a series of Soviet spacecraft: Sputnik 1, launched 4 October 1957, was the world's first artificial satellite; Sputnik 2, the world's second satellite, was launched 3 November 1957 carrying the dog Laika.

SQUINT

A term describing the angular offset between the geometrical axis (**boresight**) of an **antenna** and its radio-frequency (RF) boresight. The existence of a squint angle may be accidental—due to poor manufacturing standards or damage, which can lead to inaccuracies in **pointing**—or intentional, as part of the antenna design.

SRB

See **solid rocket booster**.

SRM

An acronym for solid rocket motor. See **rocket motor**.

SSM

See **second-surface mirror**.

SSME

See **Space Shuttle main engine**.

SSPA

See **solid state power amplifier**.

SSTDMA

See **satellite-switched time division multiple access**.

SSTO

See **single stage to orbit**.

SSUS

An acronym for **spinning solid upper stage**. See **payload assist module (PAM)**.

STABILISATION

See **three-axis-stabilised**, **spin-stabilised**, **attitude control**, **gravity gradient stabilisation**, **magneto-torquer**.

STABLE PLATFORM

See **inertial platform**.

STACK

Jargon: an assembly of **stage**s which, when stacked together, constitute a **launch vehicle** (e.g. usage: American **astronaut**s, during launch, are said to be 'riding the stack').

STACKING

The process of assembling the **stage**s of a **launch vehicle** one on top of the other. See **stack**.

STAGE

(i) A propulsive segment of a **launch vehicle**; a self-contained sub-assembly of a **multi-stage rocket** [figure S19].

Most practical launch vehicles used to deliver **payload**s to **orbit** (as opposed to **sub-orbital** trajectories) have more than one stage, each with its own **engine**s and **propellant tank**s. The most common design has three stages, although both two- and four-stage rockets exist. The lowermost stage is known as the first stage, since it is the first to provide propulsion. The main reason for having several stages is that the whole mass of the launch vehicle, including empty propellant tanks and associated structure, does not have to be carried all the way to orbit, with the result that a larger payload can be carried [see **mass ratio**].

(ii) Part of an **amplifier**, TWT **collector**, etc.

[See also **staging**, **single stage to orbit**, **upper stage**, **apogee stage**, **perigee stage**, **live stage**, **dummy stage**.]

Figure S19 A number of **Atlas** and **Centaur** rocket **stage**s in their assembly area. [General Dynamics]

STAGE SEPARATION

The moment at which a used **stage** separates from the rest of a **launch vehicle**; the process of **staging**.

STAGING

The process of **jettison**ing one **stage** of a **launch vehicle** and preparing to fire the next. This may include the firing of **retro-rocket**s on the used stage to avoid a collision with the subsequent stage, the jettisoning of an **interstage**, and the firing of **ullage rocket**s on the uppermost stage in preparation for its **ignition**. Also called stage separation.

STAR SENSOR

A device used to establish a spacecraft's **attitude** relative to a number of 'fixed stars', the positions of which are stored within an onboard database. **Sensor**s with wide fields of view are used in the **transfer orbit** to provide initial orientation, and subsequently, should a **pointing** problem occur, and narrow FOV sensors are generally used for normal, in-orbit operations.
[See also **sun sensor**, **earth sensor**.]

STATIC DISCHARGE

See **electrostatic discharge**.

STATIC FIRING

A test firing of a **rocket engine** or **rocket motor** during which the device is constrained from moving—usually conducted in a **test stand**. [See also **flight readiness firing**.]

STATION KEEPING

See **orbital control**.

STDN

See **space tracking and data network**.

STEERABLE ANTENNA

(i) A spacecraft **antenna** which can be steered by ground command to enable its **spot beam** to illuminate an alternative **coverage area**.

(ii) An **earth station** antenna with the capability to point at different **spacecraft**, or remain pointing at a single **satellite** with greater accuracy [see **box**, **free-drift strategy**].
[See also **antenna pointing mechanism**.]

STENNIS SPACE CENTER

See **NASA**.

STEP-ROCKET

An archaic term for a **launch vehicle** with several **stage**s; a multistage rocket.

STEP-TRACK

A type of **earth station** system for tracking satellites which moves an **antenna** in predetermined 'steps' in a direction dependent on the strength of the **signal** received. The signal is sampled at regular intervals and compared with the previous sample; the antenna continues to step in a particular direction until the signal power is lower than the previous sample, at which time it reverses its direction. The process then repeats.

STOICHIOMETRIC RATIO

The ratio of **fuel** and **oxidiser** required for complete **combustion** of both **propellant**s. True stoichiometric combustion leaves no excess of either component.

STORABLE PROPELLANT

See **propellant**.

STRAP-DOWN GYRO SYSTEM

An **inertial platform** in which the **gyroscope**s are connected directly to the vehicle structure ('strapped down') as opposed to being isolated from the structure in a set of **gimbal**s.
[See also **laser gyro**.]

STRAP-ON

A propulsive component of a **launch vehicle** attached to the **core vehicle** to augment its **thrust**, using either solid or liquid **propellant** (e.g. the American **Space Shuttle**'s SRBs [see **solid rocket booster**]).

STRATEGIC DEFENCE INITIATIVE

A United States development programme for a space-based defensive 'shield' against nuclear weapons. Colloquially termed 'Star Wars' after the film of the same name.

STRESS (ENGINEERING)

The branch of spacecraft design engineering concerned with the behaviour of a structure under mechanical loads.
[See also **load path, finite-element modelling, thrust structure**.]

STRESS–CORROSION CRACKING

The fracture of a material due to the combined action of tensile stress and a corrosive environment. Stress–corrosion is the corrosion of a metallic surface enhanced by local stresses.

STRETCHED (PROPELLANT TANK)

Jargon: extended or elongated. To **uprate** or improve a **launch vehicle** or spacecraft **propulsion system**, it is quite common to adapt an existing design by simply increasing the length of a **propellant tank**. This allows an increase in **propellant** capacity and overall **vehicle** performance while minimising the complexity of the design change. A launch vehicle **stage** with a stretched propellant tank is often called a 'stretched stage'.

STRINGER

A longitudinal structural brace used to reinforce the bodies of **launch vehicle**s and **spacecraft** and maintain the shape of a **skin** or other panel; a smaller version of the **longeron**.
[See also **airframe**.]

STRUCTURE MODEL

The version of a spacecraft built for the purposes of structural tests, and verification that all components and subsystems will fit in and about the proposed structure. The structure model undergoes vibration tests: if any component is 'out of spec' the **flight model** is altered accordingly.
[See also **thermal model, vibration facility, acoustic test chamber**.]

STS

An acronym for **space transportation system**.

SUBCARRIER

An electromagnetic wave of fixed amplitude and **frequency** which, in order to convey information through a radio transmission system, is modulated by a **signal** and in turn modulates a main **carrier** wave. Subcarriers are used for colour information and sound in a TV system and for **data** transmission, for example.
[See also **modulation**.]

SUB-ORBITAL HOP

A colloquial expression for a **sub-orbital** or **ballistic trajectory**.

SUB-ORBITAL (TRAJECTORY)

A **trajectory** that is less than an **orbit**. Usually the flight path of a **launch vehicle** which has insufficient energy to reach orbital velocity and so follows a parabolic or **ballistic trajectory**.

SUBREFLECTOR

A subsidiary reflector in a 'reflector antenna' system, interposed between the **feedhorn** and the main reflector in a so-called 'folded-optics' configuration. See, as examples, **Cassegrain reflector** and **Gregorian reflector**.
[See also **antenna**.]

SUB-SATELLITE POINT

The point of intersection, on the surface of a planet, of a line drawn between a satellite in **orbit** and the planet's centre.

SUBSONIC FLOW

See **mach number**.

SUBSTITUTION HEATER

See **simulation heater**.

SUBSYSTEM

Literally, a constituent part of a **system**. Although 'system' and 'subsystem' tend to be used interchangeably, 'system' is the more general term. For example, a **communications system** would include all **hardware**, **software** and procedures used to operate the system (both **space segment** and **ground segment**); a 'communications subsystem' (e.g. on a **communications satellite**) would include only the 'hardware' on board the satellite—it would be synonymous with **communications payload**.
[See also **propulsion system, propulsion subsystem**.]

SUN SENSOR

A device used to establish a spacecraft's **attitude** relative to the Sun. **Sensor**s with wide fields of view are used in the **transfer orbit** to provide initial orientation, and subsequently should a **pointing** problem occur, and narrow FOV sensors are generally used for normal, in-orbit operations.
   A common direction-sensing technique involves an arrangement of sensors grouped around the pointing axis so that they each receive the same illumination when the sensor is exactly aligned with the target source. When the sensor is directly in line with the Sun the output from each of its four cells is the same; if the sensor is tilted with respect to the Sun the output changes accordingly. This type of differential output can be used to control a satellite's **reaction wheel**s, **reaction control thruster**s, etc. This is described as a 'null-seeking' technique.
[See also **earth sensor**.]

SUN-SYNCHRONOUS ORBIT

See **heliosynchronous orbit**.

SUNSHIELD

(i) A shield used to protect a **spacecraft** in the American Space Shuttle's

payload bay from the heating effects of the Sun. The shield is mounted on the **cradle** carrying the spacecraft and operates like a 'clamshell', its two doors rotating back to allow the spacecraft to leave the payload bay [see figure P4].

(ii) Any device on a spacecraft positioned to avoid sunlight falling on sensors or other instruments.

SUPERSONIC FLOW

See **mach number**.

SURFACE COATINGS (of a SPACECRAFT)

Any of a variety of paints or 'finishes' applied to the surface of a spacecraft as part of its **thermal control subsystem**.

In a definitive sense there are four main types of surface available to the thermal design engineer, which assist in the maintenance of the spacecraft temperature between certain limits. The solar reflector and solar absorber respectively reflect and absorb most of the incident energy in the **solar spectrum**. The 'flat' reflector and the 'flat' absorber reflect and absorb practically all the incident energy, both within and beyond the solar spectrum—they have a 'flat' response across the frequency band.

[See also **second-surface mirror, thermal insulation**.]

SWING-ARM

A projection from a launch vehicle **service structure** which can be rotated towards the vehicle for access, **propellant** loading, etc, and swung away from the vehicle at the moment of **launch**, when any **umbilical**s are automatically detached. Also known as an **access arm**.

SYLDA

A French acronym for SYsteme de Lancement Double Ariane (Ariane dual-launch system) [see figure S20]. A container mounted inside the **shroud** of the **Ariane** 1 and Ariane 3 launch vehicles to allow two independent spacecraft—one within the SYLDA, the other on top—to be carried and launched by a single vehicle (the Ariane 2 launched only single spacecraft).

A similar structure on the **Titan** III launch vehicle is called an 'extension module'.

[See also **SPELDA, SPELTRA**.]

SYNCOM

The name (an abbreviation of 'synchronous communications') of a series of American **communications satellite**s which were the first to attain **geostationary orbit** (in 1963). A series designated Syncom IV (or 'Leasat') were launched in the 1980s by the American **Space Shuttle** [see figure F2].

Figure S20 Ariane **shroud** and **SYLDA** (systeme de lancement double Ariane—Ariane dual launch system). [Aerospatiale]

SYNODIC PERIOD

An orbital period defined in the **frame of reference** of the **celestial body** being orbited. See **orbit**.

SYSTEM

In general, a group or combination of interrelated or interacting elements forming a collective entity.

In space technology, along with other branches of technology, an assembly of electronic, electrical and/or mechanical components ['**hardware**'] forming a self-contained unit [e.g. **communications system, attitude and orbital control system, propulsion system, thermal protection system, life support system, space transportation system**]; any coordinated assemblage of facts, concepts, **data**, etc [e.g. **software**].

[See also **control system, subsystem**.]

T

T-MINUS...

A verbal announcement preceding the launch of a **launch vehicle** indicating the time remaining prior to **lift-off**, where 'T' refers to the time of lift-off (e.g. 'T minus 30 seconds and counting...'; also written $T-30$ s). The time after lift-off is indicated by 'T plus...etc'.
[See also **countdown, countdown sequencer, 'hold', launch window**.]

TAIL SERVICE MAST

A support structure on a **launch pad**, which carries **propellant** and/or electrical supplies to a **launch vehicle**. A simple service mast may carry a single **umbilical**, which is connected to the vehicle until the moment of launch and then ejected and carried clear of the vehicle by the mast. More complex masts are those mounted on the American **Space Shuttle**'s **mobile launch platform**, which supply **liquid oxygen** and **liquid hydrogen** to the vehicle's **external tank** [see figure S3]. At launch, they retract the umbilicals and close over them to protect them from the engine exhaust.
[See also **swing-arm, service structure**.]

TAIYUAN

A Chinese **launch site**, in Shanxi Province south of Beijing, which became operational in 1987. The first **satellite** to be launched from the site (in September 1988) was Fengyun 1, a **meteorological satellite**.
[See also **Jiuquan, Xichang**.]

TANEGASHIMA

An island off the south coast of Kyushu, Japan (at approximately 30.5°N, 131°E); the location of **NASDA**'s Tanegashima Space Center, which includes the Takesaki and Osaki **launch sites**, used respectively for small and large **launch vehicle**s. The latter has been used from the mid 1970s, mainly for the **N** and **H** series launch vehicles.

TANK

See **propellant tank**.

TANKER

A colloquial name given to space vehicles which carry supplies to an orbiting **space station**, particularly the 'Progress tankers' used to supply the Soviet Union's **Mir** station.

TAPE WRAPPING

A method of manufacturing low-density spacecraft structures, which involves wrapping a continuous length of '**pre-preg**' tape around a former to produce a hollow body, such as a propellant tank. Tape wrapping is a coarser version of **filament winding**.

TC

An abbreviation for **telecommand**.

TC&R

See **telemetry, command and ranging**.

TDM

See **time division multiplexing**.

TDMA

See **time division multiple access**.

TDRS

An acronym for tracking and data relay satellite [see figure T1].

TDRSS

An acronym for tracking and data relay satellite system. The system is intended to comprise one TDRS satellite over the Pacific Ocean (171°W), referred to as TDRSS-West, one over Brazil (41°W), referred to as TDRSS-East and an **in-orbit spare** at 79°W. It is intended to provide communications links between the ground and satellites in **low Earth orbit**s from 1200–5000 km in altitude.

TELECOMMAND

An instruction transmitted to a spacecraft from a controlling **earth station** or **mission control centre**. See **telemetry, tracking and command (TT&C)**.

TELECOMMUNICATIONS

The science and technology of **communications** at a distance, by means of **telephony, television, data communications**, etc, using the **radio spectrum, transmission lines** or **optical fibre** as the medium.

Figure T1 A tracking and data relay satellite (TDRS) and a smaller **spin-stabilised** satellite in a **payload canister** ready for loading into the **Space Shuttle**. An **inertial upper stage** (IUS) is visible below the TDRS spacecraft. [NASA]

TELECONFERENCE

A meeting between two physically separated parties whereby audio signals are transmitted between them to simulate the audio presence of the other party. At its simplest it is little more than a two-way telephone call, although it is usually held between more than two people. The better systems use wider **bandwidth** audio channels and feature background noise-reducing circuits, amongst other things.
[See also **videoconference**.]

TELEMETRY

Information or measurements transmitted by **radio-frequency** waves from a remote source to an indicating or recording device (e.g. from a spacecraft to an **earth station**). A telemetry **data** stream is usually transmitted separately from any other communications **channel**, since it concerns only the status of the spacecraft subsystems. The data are used both to control the spacecraft and establish the performance of the subsystem equipment. Historically, telemetry has been allocated its own **frequency band** and been transmitted by dedicated subsystem equipment and a **TT&C antenna** on the spacecraft, but during the transfer orbit phase it may be transmitted by part of a **communications payload** using frequencies within the payload band.
[See also **telemetry, tracking and command (TT&C).**]

TELEMETRY, COMMAND AND RANGING (TC&R)

See **telemetry, tracking and command (TT&C).**

TELEMETRY, TRACKING AND COMMAND (TT&C)

A spacecraft **subsystem** incorporating the three functions: **telemetry**, which is **downlink**ed from the spacecraft to the ground; **tracking**, whereby an **earth station** tracks the telemetry **carrier** or a **beacon** carried on the spacecraft; and **command**, whereby instructions are **uplink**ed from the ground to the spacecraft. Sometimes called a telemetry, command and ranging subsystem (TC&R).

TELEPHONY

A system of **telecommunications** predominantly for the transmission of speech.

TELEPRESENCE

Presence at a distance: a technique which allows a task or operation to be undertaken from a remote position. It generally involves the use of real-time video communications and a form of remote manipulator system, which allow an operator to carry out the task from a remote position, in relative comfort and safety. For example, an operation may be conducted in space or on a planetary surface by an operator in a **spacecraft, space station** or on Earth.
[See also **remote manipulator system**].

TELESCOPE

A device for collecting, focusing and detecting electromagnetic radiation from astronomical sources.

[See also **space telescope, Newtonian telescope, Cassegrain reflector, Gregorian reflector, Schmidt telescope, Coudé focus**.]

TELETEXT

A system of data **broadcasting** which uses the spare lines of the field blanking interval, the 'space' in the TV waveform between successive TV pictures. Bursts of digital signals carry a computer-based **videotext** service of written or graphical information to be transmitted to the consumer with the TV signal. To make use of the signals a separate decoder within the TV set is required.

Teletext is, in effect, broadcast videotext and is a non-interactive system. Viewdata, on the other hand is interactive videotext, whereby the consumer is linked to the computer by telephone and can select information as required.

TELEVIDEO

The system or process concerned with video communication (including an audio link) over a distance, between a small number of participants (typically two). The system used by a 'videophone'. The term may be used to describe a small-scale **videoconference**.

TELEVISION

The system or process concerned with transforming a visual scene, usually with an accompanying audio signal, into an electrical waveform which can be transmitted over a distance to, and displayed on, a cathode ray tube within a TV receiver.

TELEX

An international telegraphic system whereby text is transmitted between teleprinters via **transmission line**s, **microwave link**s, **satellite communications** links, etc.

TELSTAR

The name given to a series of **communications satellite**s. Telstar 1, launched into a **low Earth orbit** in-1962, was the first satellite to carry television across the Atlantic Ocean.

TERRESTRIAL TAIL

A link between a satellite **earth station** and the local terrestrial **telecommunications** network; also known as a 'backhaul link'.
[See also **single hop, double hop**.]

TEST STAND

A ground-based facility for testing **rocket engine**s and **rocket motor**s. Also called a proving stand [see frontispiece].

TETHER

A safety line which links an **astronaut** on an EVA (**extra-vehicular activity**) to the **spacecraft**. In the early days of **spaceflight** (e.g. on 'spacewalks' made from the **Gemini** capsules) the tether incorporated an **umbilical** which supplied oxygen and cooling water from the spacecraft; nowadays astronauts wear a **portable life support system** 'back-pack' and the tether is simply a safety line. EVAs conducted using the American **Space Shuttle**-based **manned manoeuvring unit** (MMU) are termed 'untethered EVAs'.
[See also **tethered satellite**.]

TETHERED SATELLITE

A satellite attached to a larger orbiting spacecraft by a metal wire or **tether** and using **gravity gradient stabilisation** to align the tether with the local vertical.

THERMAL BALANCE

See **thermal energy balance**.

THERMAL BARRIER

See **thermal insulation**.

THERMAL BLANKET

See **thermal insulation**.

THERMAL CONTROL SUBSYSTEM

The collection of equipment on a spacecraft which maintains its temperature within the limits dictated by the design.

Methods of thermal control are either passive or active. Passive control is the simplest and usually cheapest method and requires no intervention, whether man- or machine-derived, in its normal operation. It includes the use of reflective and absorptive **surface coatings**, **thermal insulation** and **radiator**s. Early spacecraft required only passive control techniques to maintain their temperature between certain operating limits, but as they became more complex and more powerful, active methods of thermal control evolved to accommodate the increase in complexity.

Active control involves more complex devices which are controlled either by direct ground command or by automatic systems working in the mode of 'feedback loop'. These range from the simple **heater** to the more advanced forms of variable conductance **heat pipe**.
[See also **optical solar reflector, second-surface mirror, heat sink, louvre, thermal doubler, heat spreader plate**.]

THERMAL CYCLE

The cyclic variation in temperature experienced by a spacecraft in **orbit** due, predominantly, to the relative position of the Sun and its **eclipse** by the Earth.

A spacecraft in **low Earth orbit** undergoes a large number of thermal cycles during its **lifetime** due to its short orbital period. It also experiences an eclipse on every orbit. This can have a detrimental effect on the spacecraft equipment, particularly the batteries which have a correspondingly short-period **duty cycle** (charging while in sunlight, discharging in shadow). A spacecraft in **geostationary orbit** is affected mainly by the diurnal variation of the Sun's position and the biannual eclipse. The differing aspects of thermal cycling are simulated during ground testing in a **thermal-vacuum chamber**.

THERMAL DOUBLER

A local increase in the thickness of a spacecraft structural panel used to improve the thermal conduction path and distribute heat more effectively.
[See also **heat sink, thermal control subsystem**.]

THERMAL ENERGY BALANCE

A quantitative expression of a body's thermal energy inputs and outputs.

For a spacecraft orbiting the Earth, the thermal energy balance may be expressed as:

$$Q_{sol} + Q_{int} = Q_{rad} + Q_{str}$$

where Q_{sol} is the rate of solar energy absorption, Q_{int} is the rate of internal energy generation, Q_{rad} is the rate of thermal energy radiation and Q_{str} is the rate of energy storage by the spacecraft.

If the spacecraft is in thermal equilibrium (energy input = energy output) then $Q_{str} = 0$ and the equation can be simplified to

$$Q_{sol} + Q_{int} = Q_{rad}$$

which effectively means that both the heat absorbed from the Sun and that generated by the spacecraft itself must be radiated to space in order to maintain the thermal energy balance of the spacecraft.

For spacecraft in **low Earth orbit**, heat input due to **albedo** and **earthshine** may also be important.
[See also **thermal control subsystem**.]

THERMAL GAITER

See **gaiter**.

THERMAL GREASE

See **interface filler**.

THERMAL INSULATION

A material which prevents the transmission of heat; a thermal barrier which limits both heat input and output. The thermal barriers used in spacecraft are of two main types: 'thermal blankets' and 'thermal stand-offs'.

The material colloquially termed a thermal blanket, but more correctly known as multi-layer insulation (MLI) due to its layered construction, is the material that gives a **satellite** its foil-wrapped, chocolate-box appearance [e.g. see figure S4]. The simplest examples of MLI consist of layers of mylar foil, aluminised on one side and crinkled to produce insulating voids. Kapton may be used as an alternative foil, and low conductivity foam or fibre-glass interlayers can improve the insulation. One example of a high-temperature superinsulating material, particularly used for rocket **nozzle**s, comprises stainless steel or titanium foil skins with silica fibre interlayers. Thermal blankets have many uses on and within **spacecraft**: they maintain the temperature of isolated components, insulate cryogenic propellant tanks to minimise **boil-off** and can be wrapped around even larger components, such as astronomical telescopes, to reduce thermal distortion which could otherwise render a precision instrument useless.

A 'thermal stand-off' is an insulating spacer, as opposed to a blanket or continuous mat, placed between a unit of spacecraft hardware and the panel or surface to which it is attached. This type of thermal barrier can be used to prevent heat loss from the spacecraft by isolating items such as **antenna** reflectors, which might otherwise act like **radiator** panels.
[See also **thermal control subsystem**.]

THERMAL LOUVRE

See **louvre(s)**.

THERMAL MODEL

The version of a spacecraft built to test the **thermal control subsystem** and verify that all components and subsystems will operate in the thermal environment of space. If any component is 'out of spec' the **flight model** is altered accordingly.
[See also **thermal-vacuum chamber**.]

THERMAL NOISE

A source of **noise** in a communications system which is due to the intrinsic random motion of electrons. The recognised method for reducing this noise is to cool the **receiver** or other relevant equipment. The greatest benefit is gained by cooling to cryogenic temperatures, close to **absolute zero**: at these temperatures molecular (and therefore electron) motion is reduced and thus so is the noise. Also called Johnson noise or Nyquist noise.
[See also **shot noise, noise power, noise temperature, noise figure, cryogenics**.]

THERMAL PROTECTION SYSTEM (TPS)

(i) In general, any system that provides protection to a **spacecraft** or **launch vehicle** from the effects of heat, particularly that due to friction with the Earth's atmosphere [see **aerodynamic heating, heat shield, ablation**], but also due to solar heating.

(ii) The formal name for the layer of heat-resistant **materials** applied to the outer surface of the American Space Shuttle **orbiter** [see figure T2]. About 70% of the vehicle's surface is covered with tiles composed of silica fibre: some 31 000 tiles, of varying shape and thickness, are required for each orbiter. The white tiles on the upper surfaces, designed to protect the orbiter from temperatures between 400 and 650 °C, are composed of a silica compound with added alumina oxide to improve their reflectivity. The tiles on the underside and other areas exposed to temperatures up to 1260 °C are coated with a black borosilicate glass. Each of the tiles is bonded first to a synthetic ('nomex') felt 'strain-isolator' pad, which allows expansion and contraction of the tile without cracking, and then to the aluminium **skin** of the orbiter. The **payload bay** doors and some upper wing surfaces, which are heated to less then 400 °C, are covered with coated nomex felt; the nose and wing leading edges, which experience temperatures exceeding 1260 °C, are clad in reinforced **carbon–carbon composite**.
[See also **thermal control subsystem**.].

Figure T2 A technician holding a sample of **Space Shuttle** 'reusable surface insulation', a component of the **thermal protection system** (TPS), still glowing from its exposure to temperatures comparable to those experienced on **re-entry**. The excellent **heat rejection** properties of the material allow the surface regions to radiate heat quickly while the interior is still 'white hot'. [NASA]

THERMAL STAND-OFF

An insulating device used to separate instruments, or other devices, from a spacecraft structure to avoid heat transfer. Also called a 'thermal barrier'. See **thermal insulation**.

THERMAL TILES

See **thermal protection system**.

THERMAL-VACUUM CHAMBER

A ground-based test facility which simulates the thermal and vacuum environment of space. The larger chambers admit the whole spacecraft. Thermal-vacuum tests simulate a number of **eclipse seasons** to test the operation of all on-board equipment. For most

satellite programmes a special **thermal model** version of the spacecraft is built to test the **thermal control subsystem**.
[See also **acoustic test chamber, vibration facility, anechoic chamber**.]

THERMOELECTRIC EFFECT

The phenomenon whereby a current is produced in a circuit made by two different conductors joined at two junctions which are maintained at different temperatures. Also called the Seebeck effect after its discoverer, the Russo—German physicist, Thomas Johann Seebeck (1770—1831). Used in the operation of the **radioisotope thermoelectric generator** (RTG).

The reverse effect, whereby a current produces a temperature difference, is known as the Peltier effect (as in 'Peltier cooling'); and the effect whereby heat is generated or absorbed when a current flows across a temperature gradient is called the Thomson effect (after William Thomson—see **kelvin**).

THIRD SPACE VELOCITY

See **escape velocity**.

THOR

An American single-**stage** liquid propellant intermediate range ballistic missile (IRBM) developed by the US Air Force in the 1950s (effectively as an equivalent to the US Army's **Redstone**). **Propellant: liquid oxygen/kerosene**. Various upper stages (e.g. Able and **Agena**) were added to the Thor first stage to form the Thor—Able and Thor—Agena, etc, which were used largely for launching scientific and weather satellites, etc. The Thor is no longer **operational** in these forms, but over a period of years the vehicle evolved into the three-stage Thor—Delta, which is now known simply as the **Delta**.
[See also **Jupiter**.]

THREE-AXIS-STABILISED (SPACECRAFT)

A spacecraft which maintains its stability using an **attitude control** system comprising **sensors** and **actuators**, as opposed to spin stabilisation [e.g. see figure T3].
[See also **spin-stabilised, spacecraft axes**.]

THRESHOLD

A limit or starting point (e.g. the minimum strength of a **signal**, etc, which will produce a specified response).

THROAT

The narrowest part of a rocket exhaust nozzle; the region of transition from subsonic to supersonic flow of the exhaust gases. See **nozzle**. [See also **mach number**.]

THROTTLING

The act of increasing or decreasing the flow of **propellant** to a **rocket engine** (typical usage: 'throttle up' to increase **thrust**, 'throttle down' or 'throttle back' to decrease).
[See also **thrust profile, dynamic pressure, rocket motor**.]

Figure T3 An artist's impression of the Olympus 1 **three-axis-stabilised** communications satellite. [British Aerospace]

THRUST

A force produced by any **propulsion system**: the product of the mass of gas ejected per unit time and the **exhaust velocity** ($t = \dot{m}v_0$), where t is measured in newtons (N), $\dot{m}$ in kg s^{-1} and v_0 in m s^{-1}.

For more precise determinations a pressure term may be added: $t = \dot{m}v_0 + A_e(P_e - P_a)$, where A_e is the area of the **nozzle** exit, P_e is the exit pressure and P_a is the ambient pressure.
[See also **lift, drag, reaction control thruster**.]

THRUST CONE

See **thrust structure**.

THRUST CYLINDER

See **thrust structure**.

THRUST FRAME

A structure which supports a **rocket engine** or **rocket motor** and, when it is fired, protects the rest of the **spacecraft** or **launch vehicle** by absorbing and dissipating the mechanical **load**s due to the **thrust**.
[See also **load path, thrust structure, diffusion member, propulsion bay**.]

THRUST PROFILE

A graphical representation of the **thrust** developed by a **rocket engine** or **rocket motor** throughout the period of **combustion**; a thrust against time characteristic. As an illustration, a typical thrust profile for a **Space Shuttle main engine** is as follows: ignition at $T-6.6$ s; thrust increases to 100% within 4.5 s (**lift-off** occurs at $T-0$ when the **solid rocket booster**s are ignited); at $T+25$ s the SSMEs are 'throttled down' to 84% for 11 s, then to 66% for 37 s; at about $T+74$ s they are 'throttled up' to 104% and will operate at this level until about $T+466$ s; thrust will then gradually be reduced to 65%, followed by engine cut-off at $T+522$ s (8.7 minutes). The reduction in thrust shortly after launch corresponds with the period of maximum **dynamic pressure**. The ability to 'throttle' the SSMEs also limits the vehicle's acceleration to 3g during ascent.
[See also **throttling**.]

THRUST STRUCTURE

(i) A spacecraft structure designed mainly to protect the spacecraft from the forces imparted by a **launch vehicle**.

In many satellites the basis of the **service module** is a central 'thrust tube' or 'thrust cylinder', often of corrugated or honeycomb construction, which provides stiffness in a direction parallel to the launch vehicle's thrust axis [see figure M2]. It often contains elements of the propulsion system (e.g. **apogee kick motor, propellant tank**s, **pressurant** tanks). The remainder of the satellite's primary structure is connected to the thrust structure [see **load path**]. In spin-stabilised satellites the 'thrust tube' may comprise a lower conical shell, an intermediate cylindrical shell and an upper cone.

An alternative design for a three-axis-stabilised satellite has no central thrust tube and relies on a rigid-box construction of panels. This type typically uses a **combined propulsion system** using liquid **bipropellant**s, where there is no need for a central tube structure to house the apogee motor.

(ii) As part of a **launch vehicle**, any structural component which distributes the thrust applied to the vehicle by the **rocket motor**(s) or **rocket engine**(s).

[See also **thrust frame, propulsion bay, diffusion member, airframe**.]

THRUST TUBE

See **thrust structure**.

THRUST VECTOR

(i) A mathematical quantity which defines the magnitude and direction of a **thrust**.

(ii) The axis along which the thrust acts.

[See also **vector steering**.]

THRUSTER

See **reaction control thruster**.

TILES (covering SPACE SHUTTLE)

See **thermal protection system**.

TIME DIVISION MULTIPLE ACCESS (TDMA)

A coding method for information transmission between a potentially large number of users, whereby all users utilise the same **frequency band** but transmit and receive at different times (measured in fractions of a second, and known as bursts). The satellite **transponder** can thus be accessed by multiple users who require a relatively wide **bandwidth**, one at a time but in extremely quick succession, which makes for efficient use of the available bandwidth. See **burst**.
[See also **frequency division multiple access**.]

TIME DIVISION MULTIPLEXING (TDM)

In **telecommunications**, the process in which several **signals** share the same **carrier** but at different times. Each signal is sampled in a predetermined sequence according to a closely defined repeating time schedule. In such a way, a large number of data **channels** can modulate the same RF carrier.

[See also **modulation, frequency division multiplexing**.]

TITAN

An American **launch vehicle** developed in the late 1950s as a back-up to the **Atlas** intercontinental ballistic missile (ICBM). The two-stage Titan I used the **liquid propellants liquid oxygen** and **kerosene**. The two-stage Titan II, which launched the **Gemini** spacecraft, used the storable liquid combination **nitrogen tetroxide/aerozine-50**. The three-stage Titan III has two **solid propellant** strap-on boosters; using the Centaur **upper stage**, Titans were used to launch the **Viking** spacecraft to Mars and the **Voyager**s to Jupiter and Saturn, for example.

The Titan became a commercially operated launch vehicle in the late 1980s. The **payload** capability of the commercial Titan III is over 14 tonnes to **low Earth orbit** (LEO); approximately 3 tonnes to **geostationary transfer orbit** (GTO) with a PAM-D [**payload assist module**]; and up to 5 tonnes to GTO and 2 tonnes to **geostationary orbit** (GEO) with an **inertial upper stage** (IUS).
[See also **Atlas, Delta, Scout**.]

TITANIUM (Ti)

A low-density metal, widely used (when alloyed with other metals) in the aerospace industry. Typical applications: spacecraft body-panel **face-skin**s, **antenna** reflector face-skins, structural components.
[See also **materials**.]

TM

(i) An abbreviation for **telemetry**.

(ii) An acronym for **thermal model**.

TOUCHDOWN

The moment of contact between a **spacecraft** and the surface of a **planetary body** (used equally for a planetary **probe** landing on Mars, for instance, or a manned spacecraft returning to Earth).
[See also **splashdown**.]

TPS

See **thermal protection system**.

TRACKING

The determination of a spacecraft's **orbital parameters** or the **trajectory** of a **launch vehicle** from a ground installation; a function of a spacecraft's **telemetry, tracking and command** subsystem. Using the telemetry **carrier**, ground stations can determine a spacecraft's direction (so-called 'angular tracking') and its distance. Some spacecraft carry a radio-frequency **beacon**, which can also be tracked.

The determination of distance is also known as 'ranging' (e.g. as in a telemetry, command and ranging system, TC&R). The determination of range variations is termed 'range-rate measurement'.

TRAFFIC

(i) In a **communications system**, the volume of 'messages' (telephone calls, **data** transmissions, etc) handled by the system in a given period.

(ii) A general term for the messages themselves.

[See also **capacity**.]

TRAJECTORY

The path described by a body moving in three dimensions under the influence of external forces. Typically used to describe the path of a **launch vehicle**, as a synonym for 'flight path'. Also used for a spacecraft moving in a path which is not an **orbit**: orbits tend to be 'closed curves'; trajectories tend to be 'open'.
[See also **ballistic trajectory, free-return trajectory, trans-lunar trajectory**.]

TRANSFER ORBIT

An elliptical Earth orbit used to transfer a spacecraft to a higher altitude orbit (i.e. usually from **low Earth orbit** to **geostationary orbit**).
See **geostationary transfer orbit**.
[See also **Hohmann transfer orbit**.]

TRANS-LUNAR INJECTION (TLI)

The point at which a spacecraft bound for the Moon leaves its **parking orbit** on a **trans-lunar trajectory**.

TRANS-LUNAR TRAJECTORY

A **trajectory** used to transfer spacecraft from Earth orbit to lunar orbit (used, for example, during the **Apollo** lunar programme).

TRANSMISSION LINE

A conducting entity designed to provide a controlled and protected propagation path for electrical signals and **carrier** waves, i.e. as opposed to propagation through **free space**. The two commonly used types are **waveguide** and **coaxial cable** [figure T4]; others include 'microstrip' and 'stripline'.

TRANSMIT CHAIN

A collection of electronic and **microwave** equipment which amplifies, filters and outputs a signal to a transmit antenna. In a spacecraft **communications payload** the transmit chain generally comprises an **IF amplifier, upconverter, TWTA, output filter** or **multiplexer** and associated switches and **hybrid**s [figure C2]. Whether or not to include the transmit antenna in the definition of the transmit chain is largely one of personal choice.
[See also **receive chain, amplifier chain, input section, output section.**]

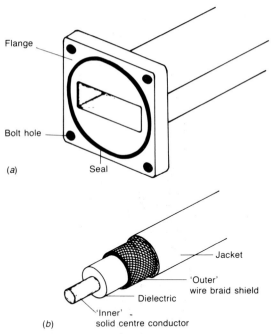

Figure T4 Two types of **transmission line**: (a) **waveguide** and (b) **coaxial cable**.

TRANSMITTER

A device which generates and amplifies a **radio-frequency** (RF) carrier, modulates the **carrier** with information and feeds it to an **antenna** for transmission.
[See also **modulation, transponder, transmit chain.**]

TRANSPONDER

Generally, a type of radio or radar transmitter–receiver that transmits signals automatically when it receives predetermined signals (derived from 'transmitter + responder'); on board a satellite, a chain of electronic communications equipment which receives, filters, amplifies and transmits a signal [see figure C2].

A spacecraft transponder comprises all the devices of the **communications payload** with the exception of the receive and transmit **antenna**s. Where there is more than one transponder chain, the alternative term 'repeater' is often used. The term 'transponder' is often used, in error, synonomously with **TWT** or **TWTA**: indeed there may be more than one TWTA in a transponder.

From the point of view of a service provider (e.g. **Intelsat, Eutelsat,** etc), a 'transponder' represents a certain amount of frequency **bandwidth** available for telecommunications services.

[See also **half-transponder, upconverter.**]

TRANSPONDER BACK-OFF

See **back-off (from saturation)**.

TRAVELLING WAVE TUBE (TWT)

A device for the high-power amplification of radio-frequency (RF) signals.

Invented in 1944 by Dr Rudolf Kompfner, an Austrian refugee working for the British Admiralty, the first practical TWT was developed at Bell Telephone Labs in 1945. TWTs have been widely used for space communications (in satellites and earth stations alike) since the early **Intelsat** and **Syncom** satellites.

The travelling wave tube (or 'tube') owes its name to its mode of operation: it is designed to cause an **RF wave** to *travel* along its length in a carefully predetermined manner. The energy for amplification is derived from a high-powered **electron beam**, which is made to interact with the RF wave carried on a **slow-wave structure**, usually in the form of a helix [see figures T5 and T6]. The interaction is such that the electrons are, on average, decelerated by the electric fields of the RF wave and thereby lose energy to the wave. The signal carried by the RF wave is therefore amplified. Individual TWTs have been built with power gains of more than 10 000 000 (70 dB).

When matched with its power supply unit (**electronic power conditioner,** EPC) the combination is referred to as a travelling wave tube amplifier (TWTA).

[See also **electron gun, collector.**]

TRAVELLING WAVE TUBE AMPLIFIER (TWTA)

An assembly of devices for the high-power amplification of **radio-frequency** (RF) signals. A TWTA comprises a **travelling wave tube** and an **electronic power conditioner** (i.e. TWTA = TWT + EPC). The acronym TWTA is sometimes pronounced 'tweeta'.

TRIAXIALITY

An attribute of an ellipsoid with three dissimilar orthogonal axes, particularly with regard to the shape of a **planetary body**; one of the mechanisms that results in the perturbation of a spacecraft's **orbit**.

The Earth, for example, is not perfectly spherical. Its rotation has caused a bulging of the equator and a flattening of the poles, its equatorial radius varies such that a cross section through the equator approximates to an ellipse, and the southern hemisphere is larger than the northern, placing the planet's **centre of mass** south of the equator.

The Earth's oblateness (the polar radius is some 21 km less than the average equatorial radius) produces perturbations in a **polar orbit**, but has little effect on **equatorial orbit**s. The difference between the minimum and maximum equatorial radii is only about 70 m, but this is sufficient to perturb a satellite in equatorial orbit: the gravitational attraction varies around its circumference, which causes the satellite to accelerate and decelerate along the orbit direction throughout the year. In fact, whereas the equatorial bulge has been known since Newton's day, the ellipticity of the equator was only detected through its effect on the orbital paths of the early artificial satellites.

[See also **perturbations**, **orbital control**, **equilibrium point**, **luni—solar gravity**, **solar wind**.]

TRI-AXIS-STABILISED

See **three-axis-stabilised**.

TT&C

See **telemetry, tracking and command**.

TT&C ANTENNA

A spacecraft **antenna** designed specifically for use with a **telemetry, tracking and command** subsystem; an **omnidirectional antenna**.

TUBE

Colloquial abbreviation for **travelling wave tube**.

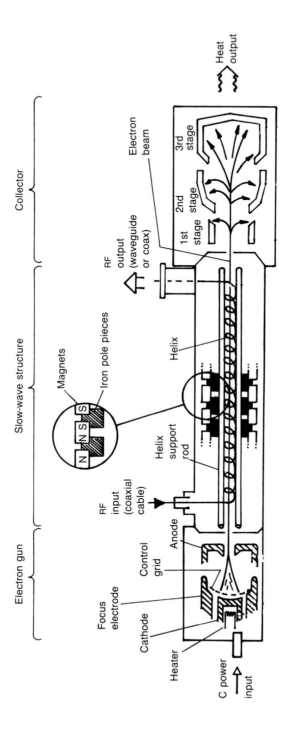

Figure T5 The component parts of a **travelling wave tube** (TWT), from left to right: the **electron gun**, **slow-wave structure** and **collector**.

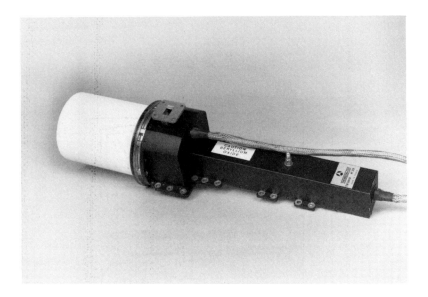

Figure T6 A high-power (230 W) **travelling wave tube** for **DBS**. The **collector** housing at left is designed to protrude from the spacecraft for direct radiation to space. [Thomson CSF]

TUMBLING

An end-over-end rotation of a body about its **centre of mass**. Some **launch vehicle** stages or components are given a tumbling motion prior to **re-entry** to discourage any aerodynamic tendencies the stage may have (which could cause it to fly a non-**ballistic trajectory**) and to make the **splashdown** point more predictable. For example, by firing a **pyrotechnic valve** in the nose cap of the American Space Shuttle **external tank**, and releasing **pressurant** gas from the **liquid oxygen** tank, it can be made to tumble at a rate of about 10 degrees/second.

TUNDRA ORBIT

An **elliptical orbit** with a 36 000 km **apogee** and a 20 000 km **perigee** (approximately). It is similar to the **Molniya orbit** in that it offers coverage of high latitudes, and similar to **geostationary orbit** in that it has period of 24 hours, which gives it one apogee as opposed to the Molniya's two. It has the advantage of no **atmospheric drag**, due to the high perigee, which also allows it to avoid the **Van Allen belts**.
[See also **orbit**.]

TURBOPUMP

A device in a liquid **propulsion system** [i.e. a **rocket engine**] which raises the pressure of the **propellant** taken from the tanks and delivers it to the engine(s) at the required pressure and flow rate. There are typically two pumps per engine, one for the **fuel** and one for the **oxidiser**. They are each driven by a turbine (hence 'turbopump') which is in turn driven by a jet of gas from a '**gas generator**'.

TV DISTRIBUTION

The transfer, via satellite, of television programming from an originating source (e.g. a TV company) to a point of distribution into the public network, or private cable network. Although such signals have been intercepted by unauthorised users, the transmissions are not intended to be received directly by the public, as they are with DBS (**direct broadcasting by satellite**).

TVRO

An acronym for television receive only. Generally used with reference to the small **antenna** installations for **direct broadcasting by satellite** (DBS).

TWEETA

Engineering slang for a TWTA (**travelling wave tube amplifier**).

TWIT

Engineering slang for a TWT (**travelling wave tube**).

TWT

See **travelling wave tube**.

TWTA

See **travelling wave tube amplifier**.

TYURATAM

The location of the **Baikonur Cosmodrome**, the leading Soviet **launch site**.

U

UCD

See **urine collection device**.

UDMH

An acronym for unsymmetrical dimethylhydrazine ($(CH_3)_2N.NH_2$), a storable liquid **fuel** used in **rocket engine**s. See **liquid propellant**.

UHF (ULTRA HIGH FREQUENCY)

See **frequency bands**.

ULLAGE

(i) The volume by which a liquid container (e.g. a **propellant** tank) falls short of being full. Also called 'ullage space'.

(ii) The quantity of liquid lost from a container by leakage or evaporation.

[See also **ullage rocket, ullage vapour**.]

ULLAGE PRESSURE

See **ullage vapour**.

ULLAGE ROCKET

A rocket **thruster** on a **liquid propellant** launch vehicle **stage**, which is fired before **ignition** of the stage's engines to position the **propellant** over the outlet valve(s). Between the **boost phase**s of successive stages the launch vehicle is **coasting** and, since there is no applied **thrust**, the propellant is 'weightless' and can float freely within its tank. Ullage rockets return the propellant to the lower end of the tank in preparation

for the next boost phase.
[See also **ullage vapour**.]

ULLAGE SPACE

See **ullage**.

ULLAGE VAPOUR

The evaporated **liquid propellant** or any other **pressurant** which occupies the 'ullage space' in a **propellant** tank. The pressure of the vapour is called the 'ullage pressure'. See **ullage**.
[See also **ullage rocket, vent and relief valve**.]

UMBILICAL

A line carrying electrical or fluid services from a service tower or other structure to a **launch vehicle**.
[See also **service structure, tail service mast**.]

UMBILICAL TOWER

See **service structure**.

UNDEREXPANDED

See **expansion ratio**.

UNFURLABLE ANTENNA

A spacecraft **antenna** which unfolds (often in the manner of an umbrella) from a position of storage once the vehicle reaches its planned **orbit** or **trajectory** [see figures U1 and T1].

Using this method, an extremely large-diameter antenna reflector can be stored in a small container. An unfurlable antenna may also have to be deployed from the spacecraft body—see **deployable antenna**.

UNIFIED PROPULSION SYSTEM

See **combined propulsion system**.

UNMANNED SPACECRAFT

Any **spacecraft** which operates without a **crew** (e.g. a **satellite** or **spaceprobe**).
[See also **manned spacecraft, man-tended spacecraft, astronomical satellite, communications satellite, meteorological satellite, reconnaissance satellite**.]

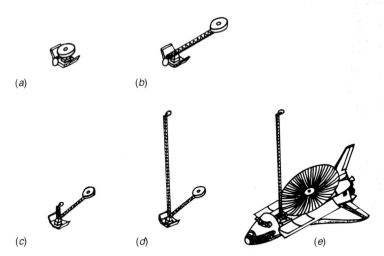

(a)   (b)

(c)   (d)   (e)

Figure U1 Deployment sequence for an **unfurlable antenna** (Antenna Technology Shuttle Experment). (a) Antenna on Shuttle pallet. (b) Deploy primary reflector canister. (c) Erect subreflector assembly. (d) Deploy subreflector. (e) Unfurl primary reflector.

UNREGULATED POWER SUPPLY

A spacecraft **power supply** which provides a variable voltage, dependent upon the power source and electrical load.

An unregulated system feeds the output of the **solar array, battery**, etc, direct to the spacecraft's **power bus**, whatever the voltage. Payload and other subsystem equipment must therefore be designed to operate across a range of voltages.

[See also **regulated power supply**.]

UNSYMMETRICAL DIMETHYLHYDRAZINE

See **UDMH**.

UPCONVERTER

A device for increasing the **frequency** of a **signal**. It contains a circuit called a **mixer**, which 'mixes' the incoming signal with a **local oscillator** frequency. This process, known as the heterodyne process or 'heterodyning', produces frequencies corresponding to the sum and the difference of the two original frequencies. The output of the upconverter is the sum signal.

A 'double-conversion' **transponder** contains a **downconverter** and an upconverter separated by a number of filter and amplifier stages.

The frequency of the output from the downconverter and the input to the upconverter is known as the **intermediate frequency** (IF). A 'single conversion' transponder contains a downconverter but no upconverter, since the conversion is directly from the **uplink** frequency to the **downlink** frequency.

UPLINK

The communications path or link between an **earth station** and a **satellite**, 'from Earth to space' [see figure U2]; also called 'feeder link'. Opposite: **downlink**.

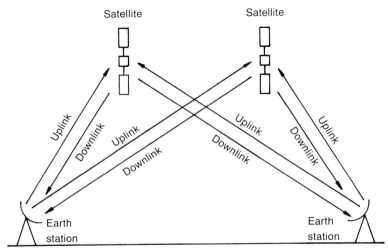

Figure U2 The **uplink** and **downlink** communications paths between **earth stations** and **satellites**.

UPPER STAGE

(i) The uppermost stage in a **multistage rocket**.

(ii) A rocket **stage** used to launch a spacecraft from the American **Space Shuttle**: e.g. the **payload assist module** (PAM), also known as a spinning solid upper stage (SSUS), and the **inertial upper stage** (IUS). (Note: PAM is also used as the uppermost stage on **Delta** and **Titan** launch vehicles.)
[See also **Centaur G/G-Prime**.]

UPRATE

To improve the **performance** of a **launch vehicle**, rocket **stage**, engine or motor. A launch vehicle can be 'uprated' by the addition of an extra stage or **strap-on** boosters, or by 'stretching' existing propellant tanks.

The 'rating' of a launch vehicle is based on its **thrust**, the **specific impulse** of its **propellant**s and, ultimately, on its **payload** capability. [See also **man-rated**.]

URINE COLLECTION DEVICE (UCD)

A bag, and its associated tubes and connectors, attached to the inside of a **spacesuit** for the hygienic collection of urine, particularly during EVA **(extra-vehicular activity)**.

V

V-2

The German World War II ballistic missile, A-4, termed the V-2 ('Vergeltungswaffe Zwei' or 'vengeance weapon two') for propaganda purposes. Its **propellant**s were a mixture of ethyl alcohol and water (**fuel**) and **liquid oxygen** (**oxidiser**). It was developed at the Peenemunde Army Experimental Station under Wernher von Braun, who surrendered to the Allies at the end of the war and became one of America's leading rocket engineers. Examples of the V-2 were taken to the USA, tested and adapted to form the beginnings of America's missile—and eventually space **launch vehicle**—industry, which produced the **Atlas** and the **Saturn V** to name but two (von Braun was instrumental in the development of the Saturn V).

VAB

See **vehicle assembly building**.

VACUUM

A region completely devoid of matter; the theoretical concept of 'nothingness'. In practice, only an approximation to a vacuum is possible, although 'outer space' represents a very good approximation for all practical purposes.
[See also **vacuum chamber**.]

VACUUM CHAMBER

A type of test chamber for spacecraft that simulates the **vacuum** properties of the space environment.
[See also **environmental chamber**.]

VAFB

See **Vandenberg Air Force Base (VAFB)**.

VAN ALLEN BELTS

Two regions of charged particles in the Earth's **magnetosphere** named after the American physicist James Alfred Van Allen (born 1914). He was instrumental in the provision of the energetic particle counters on the first American Explorer satellites, launched in 1958, which detected them. Launched later that year, Pioneer 3 recorded peaks in the particle flux at altitudes of 3218 and 16 090 km. The inner belt has since been determined to extend from about 2400 to 5600 km and the outer one from 13 000 to 19 000 km in the equatorial plane, where the belts are widest: they surround the Earth, and tail off towards the poles as they follow the magnetic lines of force that constitute the Earth's magnetic field [see figure V1]. The inner belt consists predominantly of protons (hydrogen nuclei) with energies greater than 30 MeV, being at its densest within about 35 degrees of the geomagnetic equator. The outer belt comprises mainly high-speed **electron**s (with energies from 50 keV to 5 MeV) and low-energy protons of less than 1 MeV, and spreads between about 55 and 70 degrees north and south of the geomagnetic equator.

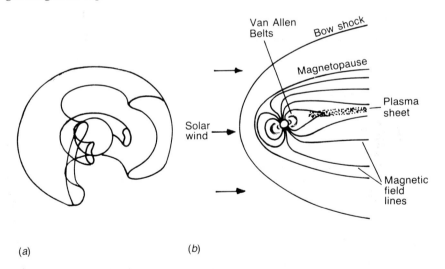

(a) (b)

Figure V1 The **Van Allen belts**. (a) The inner and outer belts around the Earth. (b) Characteristics of the Earth's magnetosphere.

The belts present a potential hazard to both manned and unmanned space missions due to their degrading effect on materials and the effects of radiation on the human body. In practice, however, most

spacecraft orbits and trajectories avoid these regions: **low Earth orbit**s tend to be below the inner belt, at altitudes of less than 1000 km; and **geostationary orbit**, where the majority of **communications satellite**s are positioned, is above the outer belt. It was decided that the **Apollo** astronauts, en route to the Moon, travelled at a velocity sufficient to reduce the radiation exposure time to safe levels.
[See also **ionisation, ionosphere, solar flare**.]

VANDENBERG AIR FORCE BASE (VAFB)

An American ICBM and space launch facility north of Los Angeles, California (at approximately 35°N, 120°W), part of which is also referred to as the 'Western Test Range' (WTR) (prior to May 1964 called the 'Pacific Missile Range'). VAFB is particularly suited for launches into **polar orbit**, since the nearest land mass due south is Antarctica. Launch facilities for the American **Space Shuttle** built at Vandenberg were 'mothballed' in 1986 following the **Challenger** accident.

VARIABLE CONDUCTANCE HEAT PIPE (VCHP)

See **heat pipe**.

VECTOR

(i) In mathematics, a variable quantity that has magnitude and direction, and can be resolved into components.

(ii) More loosely, the axis along which a force acts (e.g. **thrust vector**).

(iii) To direct a force or **thrust** (e.g. to steer a **launch vehicle**). Usage: 'to vector a **thruster**'; 'vectored thrust', etc.

VECTOR STEERING

A **launch vehicle** steering method in which one or more **rocket engine**s are **gimbal** mounted to enable the **thrust vector** to be rotated with respect to the vehicle to produce a turning moment.
[See also **vernier engine**.]

VEHICLE

In space technology, any conveyance of any type of **payload** (e.g. a **spacecraft, launch vehicle**, etc). As an alternative for 'spacecraft', the term 'vehicle' is used predominantly for **manned spacecraft**, particularly with regard to **extra-vehicular activity** (EVA), **docking**, etc.
[See also **lunar roving vehicle** (LRV), **manned manoeuvring unit** (MMU), **expendable launch vehicle** (ELV), **heavy lift vehicle** (HLV), **orbital transfer vehicle** (OTV).]

VEHICLE ASSEMBLY BUILDING (VAB)

The building at **Kennedy Space Center** in which the American **Space Shuttle** is prepared for flight. It is divided into a 'low bay' and a 'high bay', the latter being further subdivided into four bays: the low bay is used for assembling the parts of the **solid rocket boosters** (SRBs) which do not contain **propellant**; high bays 2 and 4 for the final assembly of the SRBs and preparation of the **external tank** (ET); and high bays 1 and 3 for the **integration** of the **orbiter** (transferred from the **orbiter processing facility**) with the ET and SRBs on the **mobile launch platform**, upon which the Shuttle combination will be transported to the **launch pad**.

The VAB was built in the 1960s for the **Apollo** programme and was originally designed to hold four **Saturn V** launch vehicles, one in each of the four high bays. At 158 m wide, 218 m long and 160 m tall, it remains one of the most capacious buildings in the world.
[See also **crawler**.]

VEHICLE EQUIPMENT BAY

See **instrument unit**.

VELOCITY OF LIGHT

See **light (velocity of)**.

VENERA

The name given to a series of Soviet planetary exploration **probes** launched towards the planet Venus: Venera 7 made the first radio transmissions from the surface of another planet on 15 December 1970.

VENT AND RELIEF VALVE

A device in a **liquid propellant** system, usually as part of a **launch vehicle**, which combines the functions of the vent valve and the relief valve; also called a 'boil-off valve'. A vent valve relieves the excess pressure in a propellant system when it receives an external command; a relief valve opens automatically when the pressure reaches a predetermined value. The vent function is only available before **launch** (e.g. to vent **ullage vapour** during tank-filling); the relief function is available both before and after launch.
[See also **beanie cap, ullage**.]

VENT VALVE

See **vent and relief valve**.

VERNIER ENGINE

A small **rocket engine** used to make fine adjustments to the **thrust** (magnitude and direction) of a **launch vehicle**, mounted at an angle to the main thrust axis [see figure A11]. Allows adjustments in velocity and **trajectory** to be made. An alternative method to gimballing the main engines.
[See also **gimbal**, **vector steering**.]

VERTICAL POLARISATION

See **polarisation**.

VHF (VERY HIGH FREQUENCY)

See **frequency bands**.

VIBRATION FACILITY

A ground-based test facility which simulates the vibrations transmitted to a spacecraft from a **launch vehicle**. The amplitude and frequency range of the vibrations differ between launch vehicles: the spacecraft mechanical design must avoid resonant frequencies which could damage it.
[See also **vibration table**, **acoustic test chamber**, **thermal-vacuum chamber**, **anechoic chamber**.]

VIBRATION TABLE

A test facility which subjects components or assemblies to vibrations similar to those that will be encountered during a launch. The vibration environment differs between **launch vehicle**s: data derived from actual launches can be used to test the spacecraft. Sometimes called a 'shake table' or 'shaker table'.

VIDEOCONFERENCE

A meeting between two physically separated parties whereby audio and video signals are transmitted between them to simulate the presence of the other party. The better systems use full-**bandwidth** video to produce standard **television** pictures with accompanying hi-fi audio. Inferior systems utilise reduced-bandwidth **slow-scan TV** and standard telephone-quality lines. There are also systems of intermediate quality.
[See also **teleconference**, **televideo**.]

VIDEOTEXT

A system for providing a written or graphical representation of computerised information on a television screen. When transmitted or broadcast to the consumer it is referrred to as **teletext**.

VIKING

A series of two American planetary exploration **probe**s, each comprising an 'orbiter' and a 'lander', launched towards Mars in the mid 1970s. Also the name given to a Swedish scientific spacecraft.

VLF (VERY LOW FREQUENCY)

See **frequency bands**.

VOSKHOD

One of two Soviet **manned spacecraft** launched in the mid 1960s as a follow-on to the **Vostok** programme. Since the Voskhod was little more than a 'stripped-down' Vostok, with no room for either **spacesuit**s or ejection seats but with meagre accommodation for three **cosmonaut**s, it is considered to be a spacecraft designed for political purposes. Voskhod 1, launched on 12 October 1964, carried a three-man **crew** on a 24-hour **mission** before even a two-man spacecraft had flown; and Voskhod 2, launched 18 March 1965, pre-empted the first launch of America's two-man **Gemini** by only five days. Voskhod 2 carried a crew of two leaving room for a flexible telescopic **airlock** which enabled Alexei Leonov to make the first 'spacewalk' [see **extra-vehicular activity (EVA)**].

VOSTOK

The world's first **manned spacecraft**; a series of six Soviet one-man **capsule**s which were launched into a **low Earth orbit** in the early 1960s (e.g. Vostok 1 carried the first man in space, Yuri Gagarin, into orbit on 12 April 1961 (**flight** duration: 1 h 48 min), and Vostok 6 the first woman in space, Valentina Tereshkova, on 16 June 1963 (70 h 50 min). The Vostok spacecraft comprised two main sections: a spherical **re-entry** capsule containing a rocket ejection seat; and an instrument compartment which contained service and **propulsion** subsystems (including **retro-rocket**s).
[See also **Voskhod, Soyuz**.]

VOYAGER

A series of two American planetary exploration **probe**s launched towards the outer planets of the Solar System: Jupiter, Saturn, Uranus and Neptune [see figure V2].

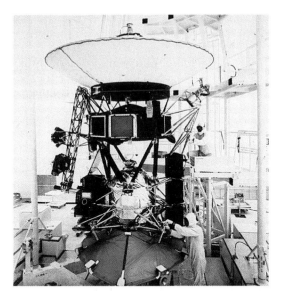

Figure V2 The **Voyager** 2 interplanetary **probe** being prepared for launch. [NASA]

W

WALLOPS FLIGHT FACILITY

See **NASA**.

WARC

An acronym for World Administrative Radio Conference, a conference convened under the auspices of the International Telecommunications Union (ITU) to plan **communications** services (particularly with regard to the allocation of radio frequencies). The first WARC for space telecommunications (WARC-ST), which made the first allocation of **frequency bands** to the **broadcasting-satellite service** (BSS), was held in 1971. A plan for the BSS in 'Region 1' and 'Region 3' [see **ITU**] was devised at WARC 1977 and ratified at WARC 1979, a plenary session which considered the whole of the **radio spectrum**; 'Region 2' was covered at the Regional Administrative Radio Conference, RARC 1983. WARC 1985 (ORB '85) and WARC 1988 discussed, amongst other things, the use of **geostationary orbit**.

WASTE MANAGEMENT SYSTEM

See **environmental control and life support system (ECLSS)**.

WAVE NUMBER

The reciprocal of **wavelength**. The number of waves per unit distance in the direction of propagation.

WAVEGUIDE

A metal tube, usually of rectangular cross section, which confines and conveys (or 'guides') electromagnetic waves.

Capable of handling higher powers than the other common form of

transmission line, coaxial cable, waveguide is therefore commonly used on the RF output side of a **high-power amplfier** (HPA), e.g. a **travelling wave tube** (TWT).
[See also **waveguide switch, feed horn, electromagnetic spectrum.**]

WAVEGUIDE SWITCH

A device designed to isolate part of a circuit and divert a signal carried in **waveguide** (one of two types of commonly used electrical **transmission line**). Devices are either switched magnetically using solenoids, causing a mechanical deviation of the waveguide run, or electronically using diodes.
[See also **coaxial cable, coaxial switch.**]

WAVELENGTH

The distance, measured in the direction of propagation, between two points of the same phase in consecutive cycles of an electromagnetic wave. Wavelength is measured in metres (m) or multiples and submultiples of metres. Common submultiples used in technical circles (apart from cm and mm) are the micron (1 μ = 10^{-6} m) and the angstrom (1 Å = 10^{-10} m). Much less common is the millimicron (1mμ = 10^{-9} m). The reciprocal of wavelength is wave number.
[See also **frequency.**]

WEB

The thickness of **solid propellant** in a **rocket motor** from the initial ignition surface to the insulated wall of the **motor case**; the total length of an end-burning charge [see **propellant grain**]. (Note: not to be confused with **shear web.**)

WEIGHTLESSNESS

The state of having little or no weight experienced in an orbiting **spacecraft** or an aircraft flying on a parabolic arc; alternatively called 'zero gravity'; sometimes expressed as 'free fall'.

Weightlessness can be experienced in **orbit** because both the spacecraft and its occupants are in free fall in the Earth's gravitational field: since both are falling at the same rate, the spacecraft offers no resistance to the falling motion of the occupant. The effect can be reproduced in an aircraft flying a **ballistic trajectory**: the aircraft is put into a dive to increase its velocity then 'pulled up', and 'pushed over' a parabolic arc. Using this technique in a light aircraft gives just a few seconds of 'zero gravity', but this can be increased to 15–45 s in a subsonic jet, about a minute in a supersonic jet and several minutes in a

rocket-powered aircraft (e.g. the **X-15**). **NASA**, for instance, generally conducts its zero-gravity simulations using a KC-135 aircraft, a modified version of the Boeing 707 which provides about 30 seconds of weightlessness per parabola. Alternatively, several seconds of 'zero-g' in a vacuum can be experienced by objects released into an evacuated 'drop tower'.
[See also **gravity**, **low gravity**, **microgravity**.]

WESTERN TEST RANGE

See **Vandenberg Air Force Base (VAFB)**.

WHITE ROOM

An air-conditioned **clean room** mounted on an **access arm** of a launch vehicle **service structure**, used to transfer a **crew** to a **manned spacecraft**.

WHITE SANDS TEST FACILITY

See **NASA**.

WIDEBAND

A modifier indicating the wide **bandwidth** of the quantity under consideration (wideband communications, wideband **signal**, **wideband noise**, etc). Essentially the same as **broadband** .

WIDEBAND NOISE

See **broadband noise**.

WIND-SHEAR

A rapid change in wind direction with height, particularly important for a **launch vehicle** rising through the atmosphere since it places increased structural (dynamic) **load**s on the vehicle.

WINDMILL TORQUE

The rotation of a **three-axis-stabilised** spacecraft about an axis in the **roll**–**yaw** plane [see **spacecraft axes**]. This can be caused by the undesirable impingement of a **thruster** plume on a **solar array** surface [see **plume impingement**], or by the intentional use of the **solar wind** in controlling the **attitude** of the spacecraft [see **solar sailing**].

WINDOW

See **launch window**.

WIRE ANTENNA

See **antenna**.

WORLD ADMINISTRATIVE RADIO CONFERENCE

See **WARC**.

X

X-1

A rocket-propelled research aircraft conceived in 1943 and designed to reach and exceed mach 1, the speed of sound, in horizontal flight: on 14 October 1947 Charles E (Chuck) Yeager broke the sound barrier. Three X-1s, dropped from the modified bomb-bay of a B-29 bomber, made 156 flights between them, setting a speed record of 1540 km h⁻¹ (about mach 1.45) and an altitude record of 21 916 m.
[See also **X-15, mach number.**]

X-15

A rocket-propelled research aircraft designed to investigate the effects of high speed at high altitude on future aerodynamic space vehicles (as opposed to ballistic **re-entry** vehicles). The main engine burnt the **propellant**s anhydrous ammonia and **liquid oxygen** and **hydrogen peroxide** was used for the **attitude control** thrusters.

The X-15s were dropped from beneath the wing of a converted B-52 bomber in a series of 199 flights (the first and last powered flights were in September 1959 and October 1968, respectively). Designed to reach speeds of up to mach 6 in horizontal flight, the X-15 set a record of 6605 km h⁻¹ (about mach 6.2), and an altitude record of 107 960 m. At one time there were plans for the X-15 to be flown into **orbit**, but this became the goal of the **Mercury** programme instead. The X-15 did, however, lead to winged **lifting body** research and eventually the **Space Shuttle**.
[See also **X-1, mach number.**]

X-BAND

8–12.5 GHz—see **frequency bands**.

XICHANG

A Chinese **launch site**, formerly known in the West as Chengdu (or Ch'eng-tu—alternative spelling) after a nearby town at approximately 28°N, 103°E, from which the **Long March** vehicle is launched.
[See also **Jiuquan**, **Taiyuan**.]

XPD

Abbreviation for cross-polar discrimination—see **discrimination**.

Y

YAW

A rotation about the yaw axis—see **spacecraft axes**.

YAW AXIS

See **spacecraft axes**.

YO-YO

A mass on a **tether** released from a spinning launch vehicle **stage** to increase its moment of inertia and thus reduce its rotation rate (usually prior to the release of its satellite **payload**).

Z

ZERO GRAVITY

A commonly used alternative term for **weightlessness**; a colloquial term referring to the conditions experienced inside a space vehicle in free fall (when the force of gravitational attraction is apparently zero). The acceleration due to gravity in a freely orbiting spacecraft is zero for most practical purposes, but this is only an approximation, since any force acting from outside or within the spacecraft will produce a gravitational acceleration—see **microgravity**.
[See also **gravity, gravitational anomaly, perturbations**.]

ZERO-MOMENTUM SYSTEM

See **reaction wheel**.

ZERO PRE-BREATHE SUIT

See **pre-breathe**.

ZONE BEAM

A **beam** formed by a **satellite** communications **antenna** which provides coverage of an area smaller than a hemisphere (see **hemispherical beam**) but larger than that covered by a **spot beam** (e.g. a zone beam may provide coverage of the contiguous United States ('CONUS') or of Europe).
[See also **beamwidth, global beam, multiple beam**.]

ZOOM ANTENNA

A type of spacecraft **antenna** with the capability to vary its area of coverage, or **footprint** (by analogy with a 'zoom lens'). An antenna with a number of **feed** elements can be used to cover a relatively large area if all the elements are fed; feeding the centre elements only can produce a smaller **coverage area** with a higher **power flux density**—the 'zoom' effect.

Classified List of Dictionary Entries

---

1. SPACECRAFT TECHNOLOGY

AC power
Accelerometer
Access panel
Acoustic test chamber
Actuator
Aerial
AM
Amplifier
Amplifier chain
Anechoic chamber
Antenna
Antenna array
Antenna efficiency
Antenna farm
Antenna feed
Antenna module
Antenna platform
Antenna pointing mechanism (APM)
Antenna radiation pattern
AOCS
Aperture antenna
Aperture efficiency
APM
APU
Array
Array antenna
Array blanket
Array shunt regulator
Astromast
Astronomical satellite
Attenuator
Attitude
Attitude and orbital control system
   (AOCS)
Attitude control
Axial ratio

Baffle
Balance mass
Ballute
Band-pass filter
BAPTA
Barbecue roll
Battery
Battery reconditioning
Bearing and power transfer assembly

Beginning of life (BOL)
BOL
Bolometer
Bolt-cutter
Boom
Boresight
Boresighting
Box
Bus
Bus voltage

Cable-cutter
Carrier generator
CEU
Channel filter
Cleanroom
Closure panel
Coax
Coaxial cable
Coaxial switch
Cold soak
Collector
Command
Communications module
Communications package
Communications payload
Communications repeater
Communications satellite
Communications subsystem
Constant conductance heat pipe
   (CCHP)
Constellation
Control electronics unit (CEU)
Cooling system
Corner-cube reflector
Cosmos
Coupled cavity
Coverglass
Coverslip
Cross-strapping

DANDE
DC power
Dead-band
Deployable antenna
Depth of discharge (DOD)

De-spin active nutation damping
   electronics
De-spun platform
Dielectric lens
Diffusion member
Diplexer
Dipole
Directivity
Dish
Dock
Docking module
Docking port
Docking probe
Double conversion transponder
Doubler
Downconverter
Drogue
Drum-stabilised
Dry mass
Dual-gridded reflector
Duty cycle

Early Bird
Earth-lock
Earth resources satellite
Earth sensor
East—west station keeping
Eclipse season
Electron gun
Electronic power conditioner (EPC)
Elliptical antenna
EMC
Encounter
End of life (EOL)
Entry interface
Environmental chamber
EOL
EPC
Equilibrium point

Face-skin
F/D ratio
Feed
Feed(er) losses
Feedhorn
Filter
Finite-element modelling
Fixed conductance heat pipe (FCHP)
Flight hardware

Flight model
Free-drift strategy
Free-flying pallet
Frequency source
Front end
Fuel cell

GaAs FET
Gaiter
Galileo
Giotto
Gorizont
Gravity gradient stabilisation
Ground spare
Guidance (inertial)
Gyro
Gyroscope
Gyroscopic stiffness

Hard dock
Hardening (of spacecraft)
Heat exchanger
Heat pipe
Heat rejection
Heat shield
Heat sink
Heat soak
Heat spreader plate
Heater
Helix
Heterodyning
High-pass filter
High-power amplifier (HPA)
Honeycomb panel
Horizon scanner
Horizon sensor
Horn
Horn antenna
Housekeeping
HPA
Hubble Space Telescope
Hybrid

IF amplifier
Image intensification
Imaging
Imaging team
Imux

Inertia wheel
Inertial guidance
Inertial platform
In-orbit spare
Input filter
Input section
Insulation
Interface filler

Klystron

Landsat
Laser
Laser ranging retro-reflector
Lens antenna
Libration period
Lifting body
Light baffle
Line heater
LNA
Load path
Local oscillator
Louvre(s)
Low-noise amplifier (LNA)
Low-pass filter
Luna

Magnetometer
Magneto-torquer
Major axis
Mariner
Maser
Mass budget
Meteorological satellite
Metsat
Microwave lens
Minor axis
Mixer
MLI
Modem
Molniya
Momentum bias
Momentum dumping
Momentum off-loading
Momentum wheel
Monocoque
Monopulse RF sensor
Multi-layer insulation (MLI)

Multipaction
Multiplexer
Mux

Navigation satellite
Nickel—cadmium cell
Nickel—hydrogen cell
North—south station keeping
Nutation
Nutation damper

Offset feed
Omni
Omnidirectional antenna
OMT
Omux
On-station
Optical solar reflector (OSR)
Orbital control
Orbital relocation
Orthomode transducer (OMT)
OSR
Output filter
Output section

Pad
Parabolic antenna
Parallel tubes
Parametric amplifier
Paramp
Parasitic station acquisition
Passive communications satellite
Payload module
Photovoltaic cell
Pin-wheel louvre(s)
Pioneer
Platform
Plumbing
Plume impingement
Polariser
Power
Power budget
Power bus
Power supply
Primary power supply
Prime spacecraft
Probe
Propulsion module

Pumped fluid loop
Pyrotechnic cable-cutter
Pyrotechnics

QM

Radar altimeter
Radiation pattern
Radiator
Radioisotope thermoelectric
    generator (RTG)
Ranging
Reaction wheel
Receive chain
Receiver
Reconnaissance satellite
Recovery
Re-entry
Re-entry corridor
Reflector antenna
Regulated power supply
Rendezvous
Repeater
Retro-reflector
Re-visit capability
RF chamber
RF power
RF sensor
RTG

SADA
SADE
SADM
Satellite
Second-surface mirror (SSM)
Secondary power supply
SEE
Sensor
Service module
SEU
Shake table
Shear web
Shunt dump regulator
Sidelobe
Simulation heater
Simulator
Single conversion transponder
Single-event effect (SEE)

Single-event upset (SEU)
Slow-wave structure
SM
Soft dock
Solar absorber
Solar array
Solar array drive mechanism (SADM)
Solar cell
Solar generator
Solar panel
Solar reflector
Solar sailing
Solid state power amplifier (SSPA)
Spacecraft
Spacecraft axes
Spacecraft integration
Spaceprobe
Space platform
Space segment
Space telescope
Space vehicle
Specific energy
Specific power
Spin-stabilised (spacecraft)
Spin-up
Spinner
Sputnik
SSM
SSPA
Stabilisation
Stable platform
Star sensor
Station keeping
Steerable antenna
Stress (engineering)
Structure model
Subreflector
Substitution heater
Sun sensor
Surface coatings (of a spacecraft)
Syncom

Tanker
TC
TC&R
TDRS
TDRSS
Telecommand

Telemetry
Telemetry, command and ranging
  (TC&R)
Telemetry, tracking and command
  (TT&C)
Telstar
Tether
Tethered satellite
Thermal balance
Thermal barrier
Thermal blanket
Thermal control subsystem
Thermal cycle
Thermal doubler
Thermal energy balance
Thermal grease
Thermal insulation
Thermal louvre
Thermal model
Thermal stand-off
Thermal-vacuum chamber
Three-axis-stabilised (spacecraft)
Thrust cone
Thrust cylinder
Thrust structure
Thrust tube
TM
Tracking
Transmission line
Transmit chain
Transmitter
Transponder
Travelling wave tube (TWT)

Travelling wave tube amplifier
  (TWTA)
Tri-axis-stabilised
TT&C
TT&C antenna
Tube
Tweeta
Twit
TWT
TWTA

Unfurlable antenna
Unmanned spacecraft
Unregulated power supply
Upconverter

Vacuum chamber
Variable conductance heat pipe
  (VCHP)
Venera
Vibration facility
Vibration table
Viking
Voyager

Waveguide
Waveguide switch
Windmill torque
Wire antenna

Zero-momentum system
Zoom antenna

2. MATERIALS

Ablation
Aluminium (Al)

Beryllium (Be)

Carbon–carbon composite
Carbon composite
Carbon-fibre-reinforced plastic
  (CFRP)
Ceramic–matrix composite (CMC)
Ceramics
CFRP
CMC

Cold welding
Composites

Degradation (of spacecraft materials)

Fatigue
Filament winding
Flat absorber
Flat reflector
Fracture mechanics

Graphite epoxy

Invar

Kevlar composite
KRP

Materials
Metal–matrix composite (MMC)
MMC

Outgassing

Pre-preg

Refractory metal

Shear
Stress–corrosion cracking

Tape wrapping
Titanium (Ti)

3. PROPULSION TECHNOLOGY

ABM
Aerospike
AKM
Annular nozzle
Apogee boost motor (ABM)
Apogee engine
Apogee kick motor (AKM)
Apogee stage
Area expansion ratio
Ascent engine
Axial thruster

Bipropellant propulsion system
Bipropellant thruster
Blowdown system
Booster
Burn
Burn-out

Characteristic velocity
Chemical propulsion
Chilldown
Chugging
Cigarette combustion
Cold-gas thruster
Combined (bipropellant) propulsion
   system
Combustion
Combustion chamber

Delta V (ΔV)
Descent engine

EHT
Ejection velocity
Electric propulsion

Electrically heated thruster (EHT)
Electromagnetic propulsion
Electron bombardment ion thruster
Electrostatic propulsion
Electrothermal propulsion
Engine
Engine cut-off
Engine re-start (capability)
Exhaust
Exhaust nozzle
Exhaust plume
Exhaust velocity
Exit cone
Expansion nozzle
Expansion ratio

Flame inhibitor
Flow separation
Fuel budget

Gas jet

Hiphet
Hiphet thruster
Hybrid rocket
Hydrazine thruster
Hypersonic flow

Igniter
Ignition
Ignition system
Impulse
Inhibitor
Ion engine
Ion propulsion
Ion thruster

4. PROPELLANTS

5. LAUNCH VEHICLES

ELA
ELV
Energia
Engine bay
Envelope
Equipment bay
Escape tower
Europa
Expendable launch vehicle (ELV)
Extension module

Fairing
Fin
Firing room
Fixed service structure
Flame bucket
Flame deflector
Flame trench
Flight
Flight controller
Flight hardware
Flight path

Gantry
Gas generator
Gimbal

H (launch vehicle)
Heavy lift launch vehicle (HLLV)
Heavy lift vehicle (HLV)
HLV
Hold
Hold-down arm
Horus
HOTOL

ICBM
Inertial upper stage
Instrument unit
Intercontinental ballistic missile
Interstage
Intertank structure
IUS

Jettison
Jupiter

Kinetic heating

Laser gyro
Laser ring gyro
Launch
Launch campaign
Launch complex
Launch control centre (LCC)
Launch escape tower
Launch pad
Launch platform
Launch profile
Launch shroud
Launch site
Launch vehicle
Launch window
Launcher
Launcher release gear
Lift-off
Live stage
Long March
Longeron

Mass ratio
Max-Q
Maximum dynamic pressure
Mobile service structure
Multistage rocket

N (launch vehicle)
Nose cone
Nose fairing

Ogive
Optical gyroscope

Payload envelope
Payload fraction
Payload integration
Payload mass ratio
Payload shroud
Pitch-over
Plugs-out test
Pogo
Pogo suppressor
Propellant dispersal system
Propellant mass fraction
Propulsion bay
Proton
Proving stand

6. SPACE SHUTTLE

7. MANNED SPACEFLIGHT

Capcom

Capsule

Capsule communicator

Carbon dioxide absorber

Carbon dioxide narcosis

Centrifuge

Closed-ecology life support system
  (CELSS)

CO_2 scrubber

Columbus

Command and service module

Command module

Communications carrier assembly

Constant-wear garment

Cooling system (of a spacesuit)

Cosmonaut

Cosmonautics

Crew

Decompression

Decompression sickness

Dysbarism

Ebullism

ECLSS

Egress

EMU

Environmental control and life
  support system (ECLSS)

EVA

Extra-vehicular activity (EVA)

Extra-vehicular mobility unit (EMU)

Extra-vehicular pressure garment

Flotation bag

Flotation collar

Foot restraint

Freedom

Gemini

Habitable module

Hermes

Hypoxia

Igloo

Ingress

Intra-vehicular activity (IVA)

Intra-vehicular pressure garment

IVA

Kvant

LEM

Life support

Life support system

Liquid cooling and ventilation
  garment (LCVG)

Lithium hydroxide canister

Lunar 'lope'

Lunar module

Lunar roving vehicle (LRV)

Man-tended spacecraft

Manned spacecraft

Man-rated

Mercury

Mir

PLSS

Portable life support system (PLSS)

Pre-breathe

Pressure suit

Primary life support system

Radio (communications) blackout

Restraint

S-band blackout

Salyut

Shirtsleeve environment

Skylab

Sleep restraint

Slide-wire

Snoopy cap

Soyuz

Space capsule

Space station

Spacecraft cabin

Spacelab

Spacesuit

Spacewalk

Splashdown

UCD

Urine collection device (UCD)

Voskhod

Vostok

Waste management system

Zero pre-breathe suit

8. COMMUNICATIONS TECHNOLOGY

Altitude-azimuth mount
AM
Amplitude modulation (AM)
Antenna gain
AOS
Aperture blockage
Atmospheric attenuation
Attenuation
Azimuth

Back-off (from saturation)
Background noise
Backhaul link
Band (frequency)
Bandwidth
Baseband
Baud
Beacon
Beam
Beam area
Beam centre
Beam edge
Beam-hopping
Beam waveguide
Beamwidth
BER
Bit
Bit error rate
Bit rate
Blockage
Boltzmann's constant
BPSK (Binary phase shift keying)
Broadband
Broadband noise
Broadcasting
Broadcasting-satellite service (BSS)
BSS
Burst
Byte

C-band
Cable
Cable TV
Capacity
Carrier
Carrier-to-interference ratio

Carrier-to-noise-power-density ratio
    (C/N_0)
Carrier-to-noise ratio
Carrier wave
Cassegrain reflector
CATV
CDMA
Channel
Channel isolation
Channel separation
Channelisation
Circular polarisation
Coast earth station
Co-channel interference
Code division multiple access
    (CDMA)
CODEC
Coding
Co-located
Common carrier
Commonality
Communication
Communications
Communications system
Companding
Co-polar(ised)
Coverage area
Cross-polar discrimination (XPD)
Cross-polar(ised)

DAMA
Data
Data communications
Data rate
Datacoms
Datum
dB
dBc ('dB relative to carrier')
dBHz ('dB hertz')
dBi ('dB isotropic')
dBK ('dB kelvin')
DBS
dBW ('dB watts')
Decibel (dB)
Declination
Decoding

Decryption
De-emphasis
Delta modulation
Demand assignment multiple access
    (DAMA)
Demodulation
Demodulator
Descrambler
Descrambling
Direct broadcasting by satellite
    (DBS)
Discrimination
Doppler shift
Double hop
Downlink
Ducting

Earth segment
Earth station
Earth terminal
Eclipse-protected
Edge of coverage
EHF (extra high frequency)
EIRP
Electron beam
Elevation (angle)
Elliptical polarisation
Encryption
End user
Energy dispersal
ENT (equivalent noise temperature)
Equatorial mount
Equivalent isotropic radiated power
    (EIRP)
Equivalent noise temperature
Eurovision

Facsimile
Fax
FDM
FDMA
Feeder link
Figure of merit (G/T)
Fixed-satellite service (FSS)
FM
Footprint
Forward error correction (FEC)
Free-space loss

Frequency
Frequency allocation
Frequency allotment
Frequency assignment
Frequency band
Frequency bands
Frequency coordination
Frequency division multiple access
    (FDMA)
Frequency division multiplexing
    (FDM)
Frequency hopping
Frequency modulation (FM)
Frequency reuse
Frequency separation
Frequency space
Frequency spectrum
FSS

Gain
Gateway station
GHz
Global beam
Gregorian reflector
Ground segment
Ground station
G/T
Guard-band

Half-power beamwidth
Half-transponder
Head-end
Head-end unit
Hemispherical beam
Hertz
HF (high frequency)
High-power DBS
Horizontal polarisation
Hour-angle
HPBW

IF
Indoor unit
Input
Integrated services digital network
    (ISDN)
Intelsat standard earth stations
Interference

RHCP (right-hand circular
    polarisation)

S-band
Satellite communications
Satellite switching
Satellite-switched time division
    multiple access (SSTDMA)
Saturated output power
Saturation
SCPC
Scrambler
Scrambling
Service area
SHF (super high frequency)
Ship earth station
Shot noise
Signal
Signal-to-noise ratio
Single channel per carrier (SCPC)
Single-entry interference
Single hop
Slow-scan TV
SMATV
Sound-in-syncs (SIS)
Spatial frequency reuse
Spillover
Splashplate
Spot beam
Spread spectrum
Spreading loss
Spurious signal
Squint
SSTDMA
Step-track
Subcarrier

TDM
TDMA
Telecommunications
Teleconference
Telephony
Teletext
Televideo
Television
Telex
Terrestrial tail
Thermal noise
Time division multiple access
    (TDMA)
Time division multiplexing (TDM)
Traffic
Transponder back-off
TV distribution
TVRO

UHF (ultra high frequency)
Uplink

Vertical polarisation
VHF (very high frequency)
Videoconference
Videotext
VLF (very low frequency)

Wideband
Wideband noise

X-band
XPD

Zone beam

9. ORBITS

Altitude
Aphelion
Apoapsis
Apogee
Atmospheric drag

Ballistic fly-by
Ballistic trajectory

Circular orbit

Clarke orbit

Decay
De-orbit
Direct orbit
Drag compensation
Drag make-up
Drift
Drift orbit

Eccentricity
Elliptical orbit
Equatorial orbit
Escape velocity

First space velocity
Free-return trajectory

GEO
Geostationary arc
Geostationary orbit
Geostationary transfer orbit (GTO)
Geosynchronous orbit
Graveyard orbit
Gravitational attraction
Gravitational boost
Gravitational perturbation
Ground track
GSO
GTO

Heliosynchronous orbit
Hohmann transfer orbit

Inclination
Injection
Insertion

Lagrange point
LEO
Libration point
Low Earth orbit
Luni–solar gravity

MCC
Mid-course correction (MCC)
Mission control
Mission control centre (MCC)
Molniya orbit

Nominal orbital position

Orbit
Orbital debris
Orbital decay
Orbital elements
Orbital inclination
Orbital injection
Orbital parameters
Orbital period
Orbital position
Orbital slot
Orbital spacing
Orbital track

Parabolic trajectory
Parking orbit
Periapsis
Perigee
Perihelion
Perturbations (of a spacecraft orbit
    or trajectory)
Plane change
Polar orbit
Prograde orbit

Retrograde orbit

Second space velocity
Semi-major axis
Sling-shot (gravitational)
Sub-orbital 'hop'
Sub-orbital (trajectory)
Sub-satellite point
Sun-synchronous orbit

Third space velocity
Trajectory
Transfer orbit
Trans-lunar injection (TLI)
Trans-lunar trajectory
Tundra orbit

10. PHYSICS AND ASTRONOMY

Absolute temperature
Absolute zero
Acceleration due to gravity
Albedo
Astronomical unit
AU

Celestial body
Centre of gravity
Centre of mass
Centrifugal force
Centripetal force/acceleration
Cosmic radiation

Coudé focus (telescope)
Cryogen
Cryogenics

Deep space
Drag
Dynamic pressure

Earthshine
Eclipse
Ecliptic
Electromagnetic pulse (EMP)
Electromagnetic radiation
Electromagnetic spectrum
Electron
Electrostatic discharge (ESD)
EM
EMI
EMP
Energy
Equinox
ESD

Frame of reference
Free fall
Free space

G-force
Gravitational anomaly
Gravitational constant
Gravitational field
Gravity

Hard vacuum
Heliopause
Heliosphere

Image
Inertial frame
Ionisation
Ionosphere

Joule–Thomson effect

Kelvin (K)

Lift
Lifting surface

Light (velocity of)
Light year
Load
Low gravity

Magnetopause
Magnetosphere
Mascon
Meteoroid
Microgravity
Micrometeoroid

Newtonian telescope

Observatory

Parsec
Phase
Photon
Planetary albedo
Planetary body
Plasma
Plasma sheath
Precession

Radiation effects
Rankine (°R)
Residual oxygen
Resolution
Robotics
Rotational velcoity (of Earth)

Schmidt camera
Schmidt telescope
Seismometer
Sidereal period
Solar constant
Solar flare
Solar spectrum
Solar wind
Solstice
Space
Space astronomy
Space–time
Spatiotemporal
Speed of light
Static discharge
Synodic period

Telescope
Thermoelectric effect
Thrust
Thrust vector
Triaxiality

Vacuum
Van Allen belts

Vector
Velocity of light

Wavelength
Wave number
Weightlessness

Zero gravity

11. SPACE CENTRES AND ORGANISATIONS

Ames Research Center

Baikonur Cosmodrome
BNSC

Cape Canaveral
Cape Kennedy
CCIR
CCITT
Centre Spatiale Guyanais (CSG)
CEPT
CNES
Comsat
Cosmodrome
COSPAS-SARSAT
CSG

Deep space network (DSN)
DOD
Dryden Flight Research Center
DSN

Eastern Test Range
EBU
Edwards Air Force Base
ELDO
ESA
ESOC
ESRIN
ESRO
ESTEC
European Space Agency
Eurospace
Eutelsat

FCC

Goddard Space Flight Center (GSFC)
Guiana Space Centre

IAF
IFRB
IGY
Inmarsat
Intelsat
Intercosmos
International Astronautical
    Federation
International Frequency Registration
    Board
International Maritime Satellite
    Organisation
International Telecommunications
    Satellite Organisation
International Telecommunications
    Union
Intersputnik
ISRO
ISY
ITU

Jet Propulsion Laboratory
Jiuquan
Johnson Space Center (JSC)
JPL

Kennedy Space center (KSC)
Kourou
KSC

Langley Research Center
Lewis Research Center

Marshall Space Flight Center

NACA
NASA
NASDA
National Space Technology
  Laboratories
NOAA
NORAD
Northern Cosmodrome

Plesetsk
PTT

RARC

San Marco platform
Space tracking and data network

STDN
Stennis Space Center

Taiyuan
Tanegashima
Tyuratam

VAFB
Vandenberg Air Force Base (VAFB)

Wallops Flight Facility
WARC
Western Test Range
White Sands test facility
World Administrative Radio
  Conference

Xichang

12. MISCELLANEOUS

Abort
Active
Aerodynamics
Assembly

Back-up
Beam-builder
Boilerplate
Breadboard

Computer enhancement
Configuration
Consumable
Control system
Conus
Coolant

Degree of freedom
Deploy
Design driver
Design lifetime

EGSE
Experimental
Frame of reference

Glitch

Graceful degradation
Ground support equipment
GSE

Hardware

Integration

Lifetime

Mach number
Maneuver
Manoeuvre
Margin
Mass-limited
MET
MGSE
Mission
Mission elapsed time (MET)
Module
MTBF
MTTF
MTTR

Negative margin
Nominal
Non-operational

OEM
Operational
Operational lifetime
Orthogonal axis
Over-design

Parabola
Paraboloid
Passive
Payload
Performance
PGSE
Pitch
Pitch axis
Pixel
Power-limited
Pre-operational
Prime system
Processing

Qualification

Radar
RAM
Rat's nest
Raw data
Real time
Redundancy
Redundant
Reliability

Remote sensing
Residuals
Retrofit
Roll
Roll axis
ROM

SDI
Simulation
Single-point failure (SPF)
Soft-fail
Software
Space debris
Space-qualified
Space science
Space technology
Spaceflight
SPF
Strategic defence initiative
Subsystem
System

Telepresence
Threshold
Tumbling

Vehicle

Yaw
Yaw axis